MULTI-CARRIER SPREAD-SPECTRUM

MULTI-CARRIER SPREAD-SPECTRUM

Edited by

KHALED FAZEL
German Aerospace Research Establishment (DLR)

and

GERHARD P. FETTWEIS
Dresden University of Technology

KLUWER ACADEMIC PUBLISHERS
BOSTON / DORDRECHT / LONDON

A C.I.P. Catalogue record for this book is available from the Library of Congress.

ISBN 0-7923-9973-0

Published by Kluwer Academic Publishers,
P.O. Box 17, 3300 AA Dordrecht, The Netherlands.

Sold and distributed in the U.S.A. and Canada
by Kluwer Academic Publishers,
101 Philip Drive, Norwell, MA 02061, U.S.A.

In all other countries, sold and distributed
by Kluwer Academic Publishers,
P.O. Box 322, 3300 AH Dordrecht, The Netherlands.

Printed on acid-free paper

Printed in the Netherlands

TABLE OF CONTENTS

*Invited

IV SYNCHRONISATION AND CHANNEL ESTIMATION

V MC-SS FOR THE UPLINK OF A CELLULAR SYSTEM

*Invited

EDITORIAL INTRODUCTION

Khaled Fazel[1], Gerhard P. Fettweis[2]

[1]German Aerospace Research Establishment (DLR)
Institute for Communications Technology
D-82234 Oberpfaffenhofen, Germany

[2]Dresden University of Technology
Department of Electrical Engineering
D-01062 Dresden, Germany

To operate future generation multimedia communications systems high data rate transmission needs to be guaranteed with a high quality of service. For instance, the third generation cellular mobile systems should offer a high data rate up to 2 Mbit/s for video, audio, speech and data transmission [1]. In addition, the important challenge for these cellular systems will be the choice of an appropriate multiple access scheme. These trends motivated many researchers to look for multiple access systems that offer high spectral efficiency.

The technique of spread spectrum originating from the military applications may allow partly to fulfill the above requirements. Advantages of spread spectrum technique are broadly known: High immunity against multipath distortion, no need for frequency planning, high flexibility, easier variable rate transmission etc. [1-2]. A multiple access scheme based on direct sequence spread spectrum, known as direct sequence code division multiple access (DS-CDMA) relies on spreading the data stream using an assigned spreading code for each user in the time domain. The capability of minimising multiple access interference (MAI) is given by the cross-correlation properties of spreading codes. In the case of multipath propagation the capability of distinguishing one component from others in the composite received signal is offered by the auto-correlation properties of the spreading codes [2]. Therefore, the receiver, i.e. the RAKE may contain multiple correlators, each matched to a different resolvable path in the received composite signal [1]. The performance of a DS-CDMA system will depend strongly on the number of active users, the channel characteristics, and on the number of arms employed in the RAKE. The system capacity is limited by self- and multiple access interference (MAI), which results from the imperfect auto- and cross-correlation properties of spreading codes. Therefore, it will be very difficult for a DS-CDMA receiver to make full use of the received signal energy scattered in the time domain and hence, to handle a full load [2].

On the other hand, the technique of multi-carrier transmission has recently been receiving wide interest for high data rate applications. The advantages of multi-carrier transmission are known such as the robustness in the case of frequency selective fading channels, in particular the reduced signal processing complexity by equalization in the frequency domain, and in the capability of narrow-band interference rejection. The technique of multi-carrier modulation (MCM) is based on transmitting data by dividing the high rate stream into several low rate streams, and by using these sub-streams to modulate different sub-carriers. By using a large number of sub-carriers, a high immunity against multipath dispersion can be provided. The symbol duration T_s of each sub-stream will be much higher than the channel time dispersion, hence the effects of inter-symbol interference (ISI) will be minimized. Since the amount of filters and oscillators is considerable for large number of sub-carriers, an efficient digital implementation of the special form of MCM called orthogonal frequency division multiplexing (OFDM) was proposed in [3] that can be easily realised by using the discrete Fourier transform (DFT). OFDM, having densely spaced sub-carriers with overlapping spectra of modulating signal, abandons the use of steep band-pass filters to detect each sub-carrier as it is used in the frequency division multiple access schemes. Therefore, it offers a high spectral efficiency. Today, the progress in semiconductor technology enables the realization of a DFT also for higher numbers of sub-carriers (up to several thousands), which has been helping for OFDM to gain much on importance. Residual ISI in MC communications can be completely suppressed if a guard interval larger than the channel time delay is inserted to each OFDM symbol [4]. The breakthrough for MC communications happened in the 1990s as OFDM was the modulation underlying Amati´s proposal which was chosen for ADSL in the US [5], and it was chosen for the European digital audio broadcasting (DAB) standard [4]-[6]. Further prominent applications include its selection in 1995 as the modulation for the European digital terrestrial television broadcasting (DTTB) system [7-8].

The advantages and success of multi-carrier (MC) modulation and the spread spectrum (SS) technique motivated many researchers to investigate the suitability of the combination of MCM with SS, known as multi-carrier spread-spectrum (MC-SS) for cellular systems. This combination, introduced in 1993 as a multiple access scheme will allow one to benefit from the advantages of both schemes: Higher flexibility, higher spectral efficiency, simpler detection techniques, narrow band interference rejection capability, etc..

Different multiple access concepts based on the combination of MC modulation with DS-CDMA for mobile and wireless indoor communications have been introduced late 1993 [9-15]. The main differences between them are in the spreading, frequency mapping and the detection strategy.

Recently, the topic of MC-SS has received widespread interest by researchers. Judging from the response to our Call for Papers and from the number of papers at international conferences, it illustrates the necessity of organizing this first international workshop covering the general issues of MC-SS, its applications, the coding & detection algorithms, the interference rejection methods, the synchronisation & channel estimation techniques, etc..

MULTI-CARRIER SPREAD-SPECTRUM MULTIPLE-ACCESS SYSTEMS

In the following we will have a brief look at three different concepts introduced in 1993. A detailed description covering these and further concepts is given in [16], which is summarised below.

The first concept, known as MC-CDMA, is based on a serial concatenation of DS spreading with MC modulation [9-12]. The high rate DS spread data stream is MC modulated in that way that the chips of a spread data symbol are transmitted in parallel on each sub-carrier. As for DS-CDMA, a user may occupy the total bandwidth for the transmission of a single data symbol. The separation of the user´s signals is performed in the code domain. The data symbols are first multiplied with the chips of the spreading code assigned to the specific user and then serial parallel converted. This reflects that the MC-CDMA system performs the spreading in the frequency domain. Hence, this approach compared to the DS-CDMA system has an additional degree of freedom which is the mapping onto the frequency domain that allows for simple methods for signal detection in the frequency domain. This concept was proposed with OFDM for optimum use of the available bandwidth. The realisation of this concept implies a guard time between adjacent OFDM symbols to prevent ISI [10-12] or to assume that the symbol duration is significantly larger than the time dispersion of the channel [9]. Thus the number of sub-carriers N_c has to be chosen sufficiently large to guarantee frequency non selective fading on each sub-carrier. The application of orthogonal codes, such as Walsh-Hadamard codes for a synchronous system, e.g. the down-link of a cellular system guarantees the absence of multiple access interference (MAI) in an ideal channel and a minimum MAI in a real channel [10-11]. As detection technique one may use equal-gain combining, zero forcing equalisation, maximum ratio combining, minimum mean square error equalisation, iterative detection, maximum likelihood detection etc..

The second and the third concepts of the combination of MCM with SS are based on first converting the data stream onto parallel low rate sub-streams before applying the DS spreading on each sub-stream in the time domain and modulating onto each sub-carrier [13-15].

Concept II, known as MC-DS-CDMA, modulates the sub-streams on sub-carriers with a carrier spacing proportional to the inverse of the chip rate. This will guarantee the orthogonality between the spectrums of the sub-streams [13-14]. If the spreading code length is smaller or equal to the number of sub-carriers N_c, a single data symbol is not spread in the frequency, instead it is spread in time domain. Spread spectrum is obtained by modulating N_c time spread data symbols on parallel sub-carriers. This concept by using high number of sub-carriers benefits from the time diversity. However, due to the frequency non-selective fading per sub-channel, frequency diversity could only be exploited if channel coding with interleaving or sub-carrier hopping is employed or if the same information is transmitted parallel on several sub-carriers. Indeed, copying the resulting spread sequence on each sub-stream may efficiently exploit the frequency diversity of the MC system. However, this approach is equivalent to a repetition coding that reduces the data rate by a factor N_c. Furthermore, here the sub-carrier spacing might be chosen larger than the chip rate that may give a higher frequency diversity for the system [14]. This concept was investigated for an asynchronous up-link scenario. For data detection, N_c coherent (non RAKE) receivers might be used.

Concept III, known as multi-tone-CDMA (MT-CDMA) applies the same data mapping and spreading as concept II. However, its sub-carrier spacing is by a factor N_c smaller than the inverse of the chip rate [15]. Thus, the N_c parallel converted data symbols before DS-spreading fulfill the orthogonality requirements. However, after the DS spreading per sub-carrier the orthogonality condition is not kept up, hence it results in inter-carrier-interference (ICI). On the other hand, the tight sub-carrier spacing enables the use of spreading codes which are by a factor of approximately N_c longer than the spreading code of a DS-CDMA. Therefore, at the expense of higher ICI, under certain conditions the system can supply more users than the DS-CDMA system [15]. Since each sub-channel might be affected by frequency selective fading, a RAKE or more complex multi-user detectors can be needed [15]. This concept was also investigated for the asynchronous up-link case.

Since 1993, the above three schemes have been deeply studied and new alternative solutions have been proposed. An overview of the research activity in this new field shows that a multitude of activities was addressed to develop these three concepts and to derive appropriate detection strategies [16-17]. A performance comparison between these concepts through software simulations is given in [17].

However, there is still a lot room for answering open questions and deriving novel solutions in this new research topic. In particular, investigations on the implementation of MC-SS systems for cellular mobile communications or other real applications have not been yet carried out sufficiently. These open questions cover the problem of the design of a suitable MC-SS scheme for both up- and down-link, simple coding and joint detection strategies, design of a powerful channel estimation and synchronisation technique (especially for the up-link), the cellular concept i.e. design of hierarchical cells such as macro, micro, pico cells, problems of non-linearities, low cost receiver design, etc.. This book will give us the opportunity to address some of these open problems more deeply.

SCOPE OF THIS ISSUE

The aim of this issue is to edit the ensemble of articles presented during the two days of the first international workshop on MC-SS that was held April 24-25 at DLR Oberpfaffenhofen.

This issue consists of five parts, where the first part is devoted to the general aspects of MC-SS. In this part first Linder gives an overview on MC-SS by considering its relation to the general multi-user and multi-sub-channel transmission methods. First, models for general multi-user and multi-sub-channel transmission are derived by taking into account the effects of the physical channel: All individual transmission channels belonging to individual users, the individual modulation, sub-channel division and multiple access schemes for all users. Then, depending on the application the advantages and the drawbacks of the MC-CDMA systems with optimum and sub-optimum multi-user detection algorithms are analysed. Vandendorpe gives an overview on the results of the MT-CDMA detection strategies. The performance of the system with diversity reception will be analysed and the design of improved detectors for single and multi-user detection will be investigated. Rohling, Gruenheid and Brueninghaus make a comparison of the different multiple access schemes, including OFDM-FDMA, OFDM-TDMA and OFDM-CDMA for the down-link of a cellular communication system. Finally, Castro *et al.* present the results of a link and system level performance of a MC-CDMA system which has been developed within the ACTS-FRAME project for the third generation mobile communications system.

In the second part of this book results on applying MC-SS for wireless indoor, mobile, and for the return channel of a cable-TV (CATV) system are presented. Harada and Prasad analyze a new multi-code and multi-carrier hybrid transmission system for future broadband mobile communications system. The problems related to multi-carrier and multi-code transmissions are discussed and solutions are proposed. Kaiser and Fazel present a new multiple access scheme based on multi-carrier spread-spectrum. This scheme with coherent detection can be used for the up- as well for the down-link of a cellular mobile communications system. Sari suggests the use of OFDMA as a multiple access scheme with frequency hopping and diversity for the return channel of a CATV-distribution. Different types of frequency hopping and error correction schemes in this multiple access technique are discussed. An M-ary orthogonal Walsh modulation scheme for a MC-CDMA system for wireless indoor communications is proposed by Dekorsy and Kammeyer. The uplink aspects are analyzed through software simulation. The comparison of a MC-SS system with the American standard IEEE-802 for WLAN is made by Banelli *et al.* by showing that the MC-SS system outperforms the IEEE-802 scheme.

The third part of this issue is devoted to coding and detection strategies for MC-SS. Maxey and Ormondroyed analyse the performance of a convolutionally coded MC-CDMA system and interference cancellation methods in fading channels. The behaviour of different equalisation schemes with coding techniques are compared in slow and fast fading channels. An approach for a MC-SS system with RAKE receiver is given by Nahler and Fettweis. By comparing two systems, namely MC-SS and DS-SS, an analysis in the time domain is carried. It is shown that by using a RAKE for both systems a similar performance can be expected. A novel MC modulated orthogonal CDMA (MCM-OCDMA) system proposal is given by Magill. The problems related to fixed and mobile subscribers in a cellular system are discussed. It is shown that the MCM-OCDMA is a particularly well suited scheme for PCS applications with fixed subscribers. Sorger, De Broeck and Schnell present a new multiple access scheme based on MC modulation for cellular mobile radio communications. In this scheme continuous transmission is used and compared to a CDMA system no spreading code is necessary to distinguish the different user signals. Hence, there will be no MAI. However, the cost to avoid MAI is to allow for ISI that requires an equalization on the receiver side. Some aspects on wide-band MC communications for military applications are detailed by Nilsson. The behaviour of the system with no guard-time and high spreading factor (low data rate) in very noisy multipath channel is investigated. The equalisation and coding for an extended MC-CDMA system over a frequency selective fading channel is detailed by Egle, Reinhard and Lindner. The performance of an iterative detection algorithm, called Soft Block Decision Feedback Equaliser is analysed. Finally Teich, Egle, Reinhard and Lindner present in the case of multipath channel the performance of a new detection method based on reccurent neural network (RNN) structure for a MC-CDMA scheme. Contrary to other neural network approaches, the RNN has the advantage that the network size and its coefficients can be obtained from parameters which characterise the communications system.

The aspects of synchronization and channel estimation for MC transmission are treated in the fourth part of this issue. Steendam and Moeneclaey analyse the sensitivity of a MC-CDMA system in the presence of carrier phase jitter. They have shown that the phase jitter gives rise to the multi-user interference. A channel estimation algorithm in time domain is proposed by Steiner for the up-link of a MC-CDMA system for mobile radio applications. The channel estimation is based on the transmission of known pilot tones. A subspace-based joint time-delay and frequency-shift estimation algorithm for a MT-CDMA system is proposed by Eric, Obradovic and Simic. This algorithm is robust especially in addressing the near-far problem. Hoeher, Kaiser and Robertson compare the performance of a two-dimensional filter with two one-dimensional filters in the case of a pilot symbol aided channel estimation in time and frequency. A family of extended Gaussian functions with a nearly optimal localisation property for OFDM scheme is given by Roche and Siohan. Unlike classical OFDM by using these functions, the presence of guard-time and rectangular window-shaping can be abandoned. Some aspects of the duality between the MC-SS and single carrier transmission system are given by Brueninghaus and Rohling. Finally, Simic, Zejak, Dukic and Eric analyse the detection of a MC complementary SS signal in Radar and Sonor systems.

The importance of the aspects of the up-link of a cellular mobile radio system based on MC-SS implies a separate part. The last part of this issue is devoted to these aspects. A combination of MC-CDMA with slow frequency hopping for the uplink is presented by Tomba and Krzymien. Both antenna space diversity with block coding are considered. The spectral efficiency of MC joint detection CDMA for the up-link is analysed by Berens, Jung and Plechinger. Here, a macro-cellular environment is considered. Results are presented in the case of slow power control schemes. Welch and Ziemer present the simulation results of a MC-DS-CDMA system in the presence of Doppler shift. A DPSK modulation with a channel gain averaging scheme for maximum ratio combiner is considered. Similar analysis is done by Ochiai and Imai for an asynchronous MC-CDMA system. Here, the utilisation of complementary sequences as the orthogonal sequences is discussed. Finally, the performance comparison of OC-FD/DS-CDMA and MC-CDMA over Rayleigh fading channel is given by Lee, Tafazolli and Evans.

REFERENCES

[1] A. Baier, U.-C. Fiebig, W. Granzow, W. Koch, P. Teder, and J. Thielecke, "Design study for a CDMA-based third generation mobile radio system," IEEE JSAC, Vol. 12, pp. 733-734, May 1994.

[2] R. Prasad, "CDMA for Wireless Personal Communications," Artech House, Boston , London, 1996

[3] S. B. Weinstein and P. M. Ebert, "Data transmission by frequency division multiplexing using the discrete Fourier transform," IEEE-Trans. On Com. Tech. Vol. COM-19, pp. 628-634, Oct. 1971.

[4] M. Alard and Lassalle,"Principles of modulation and channel coding for digital broadcasting for mobile receivers", EBU Review, Technical No. 224, pp. 47-69. Aug. 1987.

[5] J. S. Chow, J.-C. Tu and J. M. Cioffi, "A Discrete Multitone Transceiver System for HDSL Applications", IEEE JSAC, Vol. 9, pp. 895-908, 1991

[6] ETSI, ETS 300 401, “Radio Broadcast Systems: Digital Audio Broadcasting (DAB) to mobile, portable and fixed receivers“

[7] K. Fazel, S. Kaiser, P. Robertson, and M. Ruf, “A Concept of Digital Terrestrial Television Broadcasting“, Wireless Personal Communications, Vol. 2, No 1&2 pp. 9-27, 1995.

[8] ETSI, pr ETS 300 744, “Radio broadcast system for television, sound and data services; Framing structure, channel coding and modulation for digital terrestrial television“

[9] N. Yee, J.-P. Linnartz and G. Fettweis, “Multi-carrier CDMA for indoor wireless radio networks,“ Proc. IEEE-PIMRC´93, Yokohama, Japan, pp. 103-113, Sept. 1993.

[10] K. Fazel, L. Papke, “On the performance of convolutionally coded CDMA/OFDM for mobile communications system“, Proc. IEEE-PIMRC’93, Yokohama, Japan, pp. 468-472. Sept. 1993.

[11] K. Fazel, “Performance of CDMA/OFDM for mobile communications system“, Proc. IEEE-ICUPC´93, Ottawa, Canada, pp. 975-979, Oct. 1993.

[12] A. Chouly, A. Brajal, S. Jourdan, “Orthogonal multi-carrier techniques applied to direct sequence spread spectrum CDMA system“, Proc. IEEE Globecom’93, Houston, USA, pp. 1723-1728, Nov./Dec. 1993.

[13] V. DaSilva and E.S. Sousa, “Performance of orthogonal CDMA codes for quasi-synchronous communications systems,“ IEEE-ICUPC’93, Ottawa, Canada, pp. 995-999, Oct. 1993.

[14] S. Kondo and L.B. Mistein, “On the use of multicarrier direct sequence spread spectrum systems“, IEEE-MILCOM’93, Boston, USA, pp. 52-56, Oct. 1993.

[15] L. Vandendorpe, “Multitone direct sequence CDMA system in an indoor wireless environment,“ Proc. IEEE-First Symposium of Communications and Vehicular Technology, Delft, The Netherlands, pp. 4.1.1-4.1.8, Oct. 1993.

[16] S. Kaiser, “Multi-Carrier CDMA Mobile Radio Systems “, Manuscript (PhD Thesis), to appear end 1997.

[17] R. Prasad and S. Hara, “An Overview of Multi-Carrier CDMA“, Proc. IEEE ISSSTA’96, Mainz, Germany, pp. 107-114, Sept. 1996.

ACKNOWLEDGMENTS

The editors wish to express their sincere thanks for the support of the chairmen of the different sessions of the workshop, namely Prof. J. Hagenauer from Technical University of Munich, Dr. H. Harada from University of Delft, Prof. J. Lindner from University of Ulm and Dr. L. Vandendorpe from University Catholique de Louvain. Many thanks to all invited authors that through their contributions made the workshop successful. Furthermore, many thanks to the panelists, namely Dr. A. Baier from Mannesmann Mobilfunk, Dr. D. Th. Magill from Stanford Telecommunications Inc., Dr. W. Mohr from Siemens AG and Dr. R. Schweikert from DLR that have kindly accepted our invitations. We would like also to thank S. Kaiser from DLR for his active support for the local organisation of the workshop.

This first international workshop on Multi-Carrier Spread-Spectrum could not be realised without the technical support of:

Dresden University of Technology

German Aerospace Research Establishment (DLR)

IEEE-Communications Society, German Section

Information Technology Society (ITG) within VDE

Section I

GENERAL ISSUES OF MULTI-CARRIER SPREAD-SPECTRUM

Multi-Carrier Spread Spectrum: An Attractive Special Case of General Multiuser/Multisubchannel Transmission Methods

J. Lindner
Department of Information Technology
University of Ulm, Germany

Abstract

This paper considers MC-SSMA in the context of general multiuser/multi-subchannel transmission techniques. Vector transmission models for MIMO channels are described and taken as a basis for the discussion of MC-SSMA in comparison with other techniques. For some part the paper has tutorial character.

1 Introduction

For future mobile communication systems a lot of transmission and multiple access techniques have been proposed or are under discussion. One of the latest proposals is Multi-Carrier Spread Spectrum Multiple Access (MC-SSMA), a technique which combines the advantages of OFDM (Orthogonal Frequency Division Multiplexing) [1] with the advantages of SSMA, see e.g. [2], [3], [4], [5], [6]. This paper tries to give some insight into the basic relations between MC-SSMA and many other schemes which are also suitable in a frequency selective mobile radio environment. The goal is, to have a common basis for comparison and a way to draw easily first qualitative conclusions with respect to performance.

K. Fazel and G.P. Fettweis (eds.), Multi-Carrier Spread-Spectrum, 3-12.

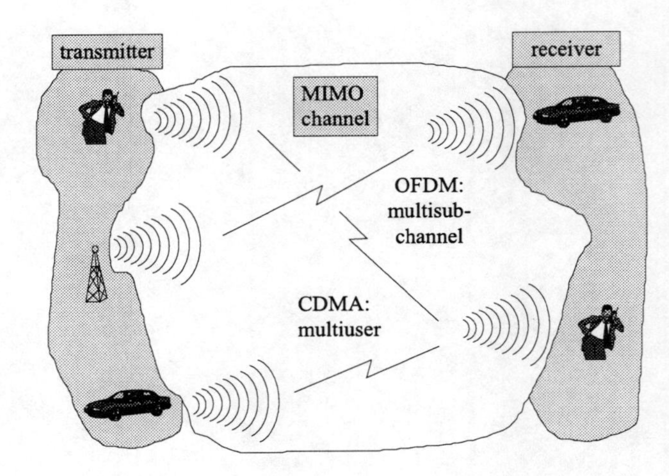

Figure 1: Multiple-input multiple-output (MIMO) channel; example: M=3 inputs, L=2 outputs

2 Multiuser/multisubchannel transmission

2.1 Scenario

Fig. 1 illustrates a simple example for the scenario assumed here. There are radio transmitters on the left and radio receivers at the right. The scenario will not be restricted to the commonly used topology with one base station and a number of mobiles, but will allow also communication between any pair of users. Yet a restriction to special cases can always be made. Because a transmission in one direction will be modeled, a number M of transmitting and a number L of receiving users will be assumed. M and L need not to be the same. In addition, fig. 1 shows, that there might be a mixed environment. Transmissions with subchannel techniques like OFDM might coexist e.g. with CDMA.

The first step towards a model for this scenario is shown in the middle part. It is a so-called *Multiple Input Multiple Output* (*MIMO*) channel. At the input of this MIMO channel there is the generalized transmitter and at the output the generalized receiver. *Generalized* means here, that all single transmitters and all single receivers are taken as a whole. This implies a possible direct exchange of information between the individual transmitters and in the same way between individual receivers. The commonly assumed separate transmitters and receivers are included as special cases. The reason for taking this general approach is, that subchannel methods as well as diversity transmission and diversity reception will be included. Subchannels usually belong to one real transmitter and a corresponding conventional receiver detects data sysm-

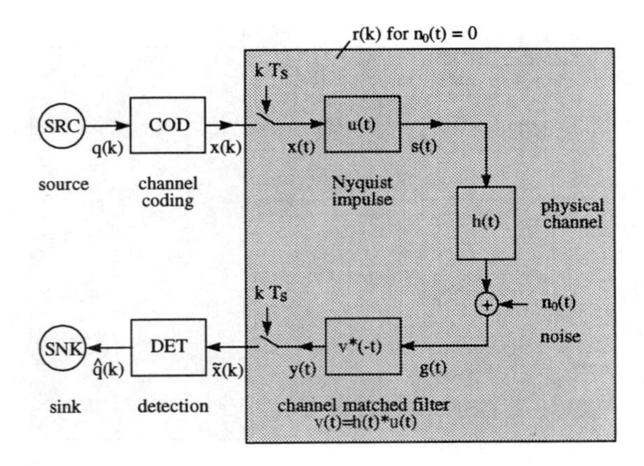

Figure 2: Continuous and discrete-time scalar transmission model

bols on individual subchannels not independently. For diversity reception the two separate receiving users in fig. 1 could be replaced by only one user who can take the two different receive signals for more reliable detectection. In a similar way transmitter diversity may be included.

From a transmission point of view the scheme in fig. 1 will be called a *general multiuser (GMU) transmission system* or a *multiuser/multisubchannel (MU/MSC) transmission system.* In the next subsection this system will be modeled by a vector-valued transmission.

2.2 Vector-valued transmission

Fig. 2 shows in its upper part a simple conventional model for a transmission with linear modulation like mPSK or mQAM. The source generates a sequence $q(k)$ of source symbols, which are fed to the channel coding. Source encoding shall be included in the source. The channel coding box puts out a sequence $x(k)$ of transmit symbols, which will in general be complex-valued, e.g. $x(k)\varepsilon\{\pm 1 \pm j\}$ in case of 4PSK. Coded modulation is therefore included in COD. The rest of the transmitter is a symbolic description of the generation of the transmit signal $s(t)$, i.e.

$$\begin{aligned} s(t) &= u(t) * x(t) \\ &= \sum_k x(k) \cdot u(t - k \cdot T_s) \end{aligned} \tag{1}$$

$u(t)$ is a Nyquist impulse, T_s the symbol interval connected with the transmission of each $x(k)$, and $*$ means convolution. The physical channel will be modeled by a linear time invariant system with impulse response $h(t)$ and additive white Gaussian noise (AWGN) with sample function $n_0(t)$ at the input

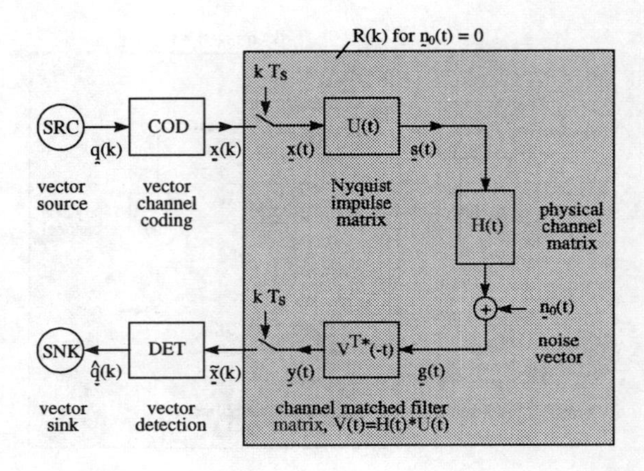

Figure 3: Continuous and discrete-time vector transmission model

of the receiver. The lower part of fig. 2 shows the corresponding receiver. It consists in its first part of a channel matched filter (CMF), which is matched to $v(t) = h(t) * u(t)$, i.e. to the concatenation of the Nyquist impulse with the channel impulse response. The symbol-spaced sampling of the CMF output signal $y(t)$ gives the sequence

$$\begin{aligned} \tilde{x}(k) &= y(k \cdot T_s) \\ &= r(k) * x(k) + \tilde{n}(k) \end{aligned} \tag{2}$$

$r(k)$ is the discrete-time channel impulse response on symbol basis. It includes the physical channel as well as the Nyquist impulse $u(t)$:

$$r(k) = v^*(-t) * v(t) \,|_{t=k \cdot T_s} \tag{3}$$

$\tilde{n}(k)$ is a sequence of noise samples taken at the output of the CMF. Its autocorrelation function is given by $r(k)$ and the two-sided spectral power density N_0 of the WGN process:

$$\varphi_{\tilde{n}\tilde{n}}(k) = N_0 \cdot r(k) \tag{4}$$

The $\tilde{x}(k)$ form a sufficient statistics with respect to maximum likelihood (ML) or maximum a posteriori (MAP) detection of the sequence of source symbols $q(k)$. The output of the detection algorithm will be denoted by $\hat{q}(k)$.

It can be shown (see e.g. [7]), that this transmission model can be generalized in a straightforward manner. Fig. 3 shows the result. Instead of scalar sequences and signals there are now vector sequences and vector signals (which is denoted by underlining). At the transmit side the components of $\underline{q}(k)$ and $\underline{x}(k)$ might correspond with sequences transmitted by individual transmitters as well as with sequences transmitted by only one transmitter over subchannels, e.g. with OFDM. Also a mix of both cases is allowed. The model does not restrict the definition of the vector components. Also transmitter diversity can be included by defining two or more of the components $x_i(k)$ to be the same symbol to be transmitted. The impulse responses $u(t)$, $h(t)$ from the scalar model are now *matrices* $U(t)$ and $H(t)$ of impulse responses. Such matrices describe linear time invariant vector or MIMO systems. If there are M users/subchannels and diversity transmissions in total, then $U(t)$ is a MxM diogonal matrix:

$$U(t) = \text{diag}(u_1(t), u_2(t), ..., u_M(t)) \tag{5}$$

The LxM matrix $H(t)$ describes the physical MIMO channel with M inputs and L outputs. One impulse response entry $h_{lm}(t)$ belongs to the transmission of the n-th user/subchannel or diversity transmit symbol over the radio channel to the l-th receiver:

$$H(t) = [h_{lm}(t)]\,; \quad l = 1, ..., L; \quad m = 1, ..., M \tag{6}$$

Asynchroneous transmitters can be modeled by a corresponding time shift included in the corresponding impulse responses $h_{lm}(t)$. The CMF from the scalar case becomes now a *CMF matrix* matched to $V(t) = H(t) * U(t)$. Eq.(2) from above will now be generalized to

$$\begin{aligned} \underline{\tilde{x}}(k) &= \underline{y}(k \cdot T_s) \\ &= R(k) * \underline{x}(k) + \underline{\tilde{n}}(k) \end{aligned} \tag{7}$$

$R(k)$ is the discrete-time *channel matrix* (actually it is a *sequence* of matrices) corresponding to $r(k)$ in the scalar case:

$$R(k) = V^{T*}(t) * V(t)\,|_{t=k \cdot T_s} \tag{8}$$

For the noise vector $\underline{\tilde{n}}(k)$ at the output of the CMF matrix in general a correlation between the components $\tilde{n}_i(k)$ and also a correlation with respect to k results. The correlation matrix can be calculated easily:

$$\Phi_{\tilde{n}\tilde{n}}(k) = N_0 \cdot R(k) \tag{9}$$

Fig. 3 illustrates also the discrete-time transmission model according to eq.(7)

2.3 Decomposition of R(k)

It is possible to decompose $R(k)$ as follows [8]:

$$R(k) = U^{T*} \cdot R_{HH}(k) \cdot U \tag{10}$$

$$R_{HH}(k) = \sum_{l=1}^{L} R_{H_l H_l}(k) \tag{11}$$

In eq.(10) only $R_{HH}(k)$ is dependent on k. Also, there is no dependence on t like before and instead of convolution there are only normal matrix products. U is constructed from $U(t)$ by replacing each entry in $U(t)$ by samples of the Nyquist impulses $u_i(t)$. They must be written in column direction with a non-overlapping column vectors. Therefore, the $u_i(t)$ need to be bandlimited to frequency f_g and the samples must then be taken with the sampling interval $\Delta t \leq \frac{1}{2f_g}$. The matrix $R_{HH}(k)$ represents the influence of the physical channel and the overlapping of vector symbols (or blocks) on the channel, resulting in *interblock interference* (*IBI*). Each term in the sum of eq.(11) can be interpreted as one of L diversity paths and it is calculated as follows:

$$R_{H_l H_l}(k) = H_l^{T*} \cdot I(k) \cdot H_l \tag{12}$$

The matrices H_l consist of M submatrices:

$$H_l = \begin{bmatrix} H_{l1} & H_{l2} & \cdot & \cdot & \cdot & H_{lM} \end{bmatrix} \tag{13}$$

These M submatrices have Toeplitz form with shifted versions of discrete-time channel impulse responses in its rows, i.e.

$$H_{lm} = \text{Toeplitz}(h_{lm}(i \cdot \Delta t)) \tag{14}$$

$h_{lm}(i \cdot \Delta t)$ are samples of the individual channel impulse response $h_{lm}(t)$ which describes the transmission from input m of the MIMO channel to output l. $I(k)$ in eq.(12) models the IBI caused by the MIMO channel. It is a *shift matrix* which shifts the the components of a column vector by $k \cdot N_s$ positions in upper direction. N_s is the symbol interval T_s counted in samples according to the sampling theorem: $N_s = \frac{T_s}{\Delta t}$. It will assumed to be an integer. $I(0) = I$ is the identity matrix and $I(k)$ has the main diagonal of I shifted by $k \cdot N_s$ positions to the right.

It must be noted that this description reduces to a conventional multiuser receiver only for $L = 1$. For $L > 1$ the diversity reception of the generalized receiver can benefit of all components $g_i(t)$ of the vector input signal $\underline{g}(t)$.

2.4 Basic structures of R(k) and detection algorithms

Fig. 4 shows some basic structures of $R(k)$ for different combinations of MU/MSC transmissions. This table is taken from [7] and it was discussed there in more detail. It is remarkable that many combinations and techniques result in similar matrices and that a MU/MSC detection algorithm does not care about the special method. It will "notice" only $R(k)$! Matrices having non-zero elements only next to the main diagonal are of advantage with respect to the detection algorithm. But diagonal matrices like in case of OFDM with zeros on the main diagonal are also not the goal. There is a degree of freedom in getting an $R(k)$ with good conditions: the matrix U, which determines the modulation and multiple access scheme, can be defined properly.

case	basic structure of $\underline{R}(k)$	interference			typical technique		user		guard band / guard time
		IBI	IUI	ISCI	sub-channel	user	sync	async	
1	k= ... -1 0 1	x	x	x	CDM	CDMA	x	x	0
2	... 0 0 ...	0	x	x	OCDM	OCDM	x		x
3	... ...	x (ISI)	0	0	FDM (conventional)	FDMA	x	x	x
4	... 0 0 ...	0	0	0	OFDM	OFDM	x		x
5	... ...	x	see text		CDM FDM TDM*	TDMA FDMA CDMA TDMA TDMA FDMA	x x x x x x	 x x x	x x x x x x
6	... 0 0 ...	0	see text		OCDM OFDM MC-CDMA MC-SSMA	TDMA OFDM OCDM 	x x x x x		x x x x x

Figure 4: Basic structures of $R(k)$ and typical techniques, IBI: interblock interference, IUI: interuser interference, ISCI: intersubchannel interference

A detection algorithm which is very well suited in this context is a *vector decision feedback equalizer* (*vector DFE*), see fig. 5. Compared to ML algorithms it is suboptimum, but with respect to complexity and performance it is a very good solution for an uncoded transmission, also all of its variants, see [9] and the references given there. In the coded case it can also be used if properly concatenated with decoding [10], [9]. Other very attractive algorithms are based on recurrent neural networks, see [11] and the references given there. In any case the performance of those algorithms with respect to its complexity, bandwidth and power efficiency depends strongly on the choice of U (or $U(t)$).

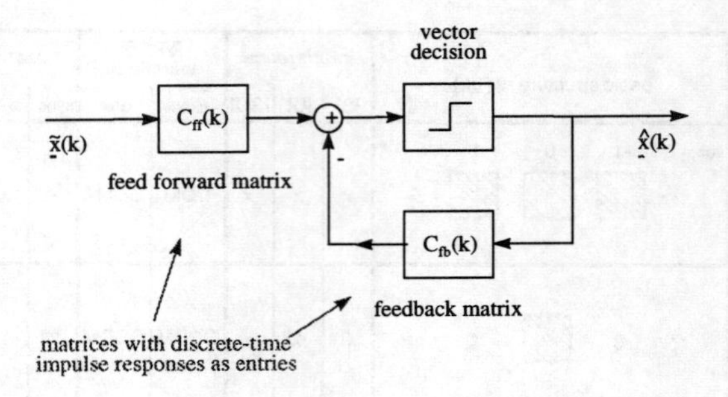

Figure 5: Detection algorithms, example: vector DFE (uncoded transmission)

3 MC-SSMA

MC-SSMA is based on MC transmissions or on OFDM. For OFDM U is the invers fourier matrix F^{-1} and hence $U^{T*} = F$. The common repitition of the transmit signal over a guard time T_G leads to a cyclic matrix $R_{HHTG}(k)$ if T_G is greater than the greatest duration of all individual channel impulse responses $h_{lm}(t)$ and if a one-period window is taken at the receiving side for detection. The windowing also leads to IBI-free transmission, i.e. $R_{HHTG}(k) = R_{HHTG}(0) \cdot \delta(k)$. With the abbreviation $R_{HHTG}(0) = R_{HHTG}$ the following results:

$$\begin{aligned} \underline{\tilde{x}}_{OFDM}(k) &= U^{T*} \cdot R_{HHTG} \cdot U \cdot \underline{x}_{OFDM}(k) \\ &= F \cdot R_{HHTG} \cdot F^{-1} \cdot \underline{x}_{OFDM}(k) \\ &= \Lambda_H \cdot \underline{x}_{OFDM}(k) \end{aligned} \tag{15}$$

Λ_H is a diagonal matrix with the squared absolute eigenvalues of R_{HHTG} on the main diagonal. The eigenvalues are identical with the squared asolute values of the transfer functions of the channels at the specific subcarrier frequency. In case of frequency selective fading some of the subcarriers may be faded out, so that zeros or small values on the main diogonal of Λ_H can result and some components in $\underline{\tilde{x}}_{OFDM}(k)$ vanish. The idea of MC-CDMA is to prevent this by spreading the energy of each data symbol over some of the subcarriers to get good performance already in the absense of coding - see also [12]. To keep the total data rate uneffected, other users or subchannels must use the same subcarriers. The separation is done like in CDM/CDMA Mathematically this means:

$$\underline{x}_{OFDM}(k) = U_0 \cdot \underline{x}_{MCCDMA}(k) \tag{16}$$

Now $\underline{x}_{MCCDMA}(k)$ is the sequence of vectors to be transmitted. U_0 is a spreading matrix in front of the inverse fourier transform. Usually it consists of Walsh

Hadamard (WH) submatrices on the main diogonal:

$$U_0 = \mathrm{diag}(WH_1, WH_2, ..., WH_M) \quad (17)$$

It is common to take identical submatrices, i.e. $WH_i = WH$. But other orthogonal matrices can also be taken - see discussion in [12]. The structure of U_0 looks like $R(k)$ in the last row of the table in fig. 4. Despreading means

$$\begin{aligned} \underline{\tilde{x}}_{MCCDMA}(k) &= U_0^{T*} \cdot \underline{\tilde{x}}_{OFDM}(k) \quad (18) \\ &= U_0^{T*} \cdot F \cdot R_{HHTG} \cdot F^{-1} \cdot U_0 \cdot \underline{x}_{MCCDMA}(k) + \underline{\tilde{n}}(k) \\ &= \left(F^{-1} \cdot U_0\right)^{T*} \cdot R_{HHTG} \cdot \left(F^{-1} \cdot U_0\right) \cdot \underline{x}_{MCCDMA}(k) + \underline{\tilde{n}}(k) \end{aligned}$$

With a new $U = F^{-1} \cdot U_0$ this can be interpreted as a conventional spreading in time domain with the fourier invers of the spreading matrix U_0. So in principle there seems to be no difference between orthogonal CDM (OCDM) and MC-CDMA - only special spreading sequences are taken for MC-CDMA. But the difference is that with MC-CDMA the spreading can adapted to the frequency selective behaviour of the channel much better in frequency domain. Then it is easier to find a good compromise between performance and complexity of the detection algorithms, see e.g. [9], [11] in these proceedings.

4 Conclusion

The advantage of MC-SSMA or MC-CDMA and its variants compared with pure OFDM is the greater robustness against frequency selective fading aready without the help of channel coding, which, of course, leads to further improvements. One disadvantage compared with OFDM is the increase in complexity, but with the adjustment of the size of the spreading submatrices and with appropriate detection algorithms one can always find a good compromize between this increase in complexity and the gain in performance. Compared with other guard time using techniques (orthogonal CDM, OCDM) the avantage of MC-CDMA is also its flexible, channel-adapted spreading and the possibility to achieve high bandwidth efficiency more easily. On the other hand, the generalized transmission model explained in this paper has shown, that derivatives with special spreading matrices U might exist, which meet given requirements even better, e.g. with respect to nonlinearities in the transmitter.

References

[1] J.A.C. Bingham. Multicarrier Modulation for Data Transmission: An Idea Whose Time has Come. *IEEE Commun. Mag.*, pages 5–14, 1990.

[2] A. Chouly, A. Brajal, and S. Jourdan. Orthogonal Multicarrier Technique applied to Direct Sequence Spread Sprectrum CDMA Systems. *Proc. IEEE Globecom, Nov. 1993*, pages 1723–1728, 1993.

[3] N. Yee, J. P. Linnartz, and G. Fettweiss. Multicarrier CDMA in Indoor Wireless Radio Networks. *PIMRC '93, Yokohama*, pages 109–113, 1993.

[4] K. Fazel. Performance of CDMA/OFDM for Mobile Communication Systems. *Proc. ICUP '93*, pages 975–979, 1993.

[5] C. Reiners and H. Rohling. Multicarrier Transmission Technique in Cellular Mobile Communication Systems. *Proc. IEEE VTC' 94*, pages 1645–1649, 1994.

[6] K. Fazel, S. Kaiser, and M. Schnell. A Flexible and High Performance Cellular Mobile Communications System Based on Orthogonal Multi-Carrier SSMA. *Wireless Personal Communications*, pages 121–144, 1995.

[7] J. Lindner. MC-CDMA and its Relation to General Multiuser/Multisubchannel Transmission Systems. *Proc. ISSSTA '96*, 1996.

[8] J. Lindner. On General Multiuser/Multisubchannel Transmission Systems. *Internal Report, Department of Information Technology, University of Ulm, Germany*, 1997.

[9] J. Egle, M. Reinhardt, and J. Lindner. Equalization and Coding for Extended MC-CDMA over Time and Frequency Selective Channels. These Proceedings, 1997.

[10] M. Reinhardt, J. Hess, and J. Lindner. Combined Decision Feedback Equalization and TCM-Decoding for Multiple Input/Multiple Output Channels with Applications in CDMA-Systems. *Proc. Third UK/Australian International Symposium on DSP for Communication Systems, University of Warwick, UK, 12-14 Dec. 1994*, 1994.

[11] W.G. Teich, J. Egle, M. Reinhardt, and J. Lindner. Detection Methods for MC-CDMA Based on a Recurrent Neural Network. These Proceedings, 1997.

[12] J. Lindner. Channel Coding and Modulation for Transmission over Multipath Channels. *AEÜ*, 49, 1995.

Overview of the results about multitone CDMA detection

L. Vandendorpe[1] *

UCL Communications and Remote Sensing Laboratory, 2, place du Levant, B-1348 Louvain-la-Neuve, Belgium. Phone : +32 10 47 23 12, Fax : +32 10 47 20 89 - E-Mail : vandendorpe@tele.ucl.ac.be

Abstract

In the present paper we review the concept of multitone CDMA. We also review the structure and performance associated with fractionally spaced receivers: linear or decision-feedback fractionally spaced multiuser detectors. The sensitivity of the detectors in near-far conditions and the robustness against timing errors is also covered.

1 Introduction

A first combination of multitone (MT) modulation with CDMA (Code Division Multiple Access) has been introduced in [9]. In this system, spreading is performed first and then the different chips are mutiplexed to enter the MT or OFDM (orthogonal frequency division multiplexing) modulator. When the number of chips per symbol is equal to the number of tones, the basis functions associated with this type of modulation are obtained by multiplying the MT or OFDM basis functions by the corresponding chip. In a multiuser scenario, the basis functions allocated to the different users are of course different because the codes are different. Besides if synchronism of the different users can be assumed, the basis functions are orthogonal at the emitter side.

* This author would like to thank the Belgian NSF for its financial support

K. Fazel and G.P. Fettweis (eds.), Multi-Carrier Spread-Spectrum, 13-21.

In [6], another MT-CDMA proposal has been issued independently. In this alternate proposal, MT modulation is performed first and then the MT symbols are spread. The motivation for considering this combination was to try to benefit from the resistance of multitone signals against multipath propagation while providing the system with a multiple access capability. Besides, CDMA implemented by means of direct sequence spreading makes it also possible to benefit from diversity reception like selection diversity or rake reception. When spreading is put on top of a multitone signal as proposed in [6], the basis functions associated with the different signals are the product of the OFDM basis functions with the code of the user under consideration. Another important important feature of the system is that for a constant bandwidth, the ratio between the number of chips and the number of tones is constant. Hence when the number of tones increases, the number of chips per symbol does as well and longer spreading sequences can be used. However the waveforms allocated to the different users are not orthogonal. The present paper deals with the proposal made in [6]. The performance of MT-CDMA (multitone CDMA) in a multiuser and multipath scenario with diversity reception (selection and rake) has been investigated in [6]. The total interference was modelled by means of the gaussian approximation. The advantage of MT-CDMA over classical CDMA has been demonstrated in that context. When a multitone CDMA (MT-CDMA) signal is transmitted over a frequency selective channel in the presence of other users, the signal experiences ISI (inter symbol interference) and ICI (inter-carrier interference) which are basically self-interference, and also MAI (multiple access interference) due to the activity of other users and the nonideal properties of the correlations between users. These interferences limit the global performance of such a system, but improved detectors can be used to build improved decision variables. Major advances in the field of joint detection or interference suppression are due to Lupas and Verdu [4]. A very good overview of the recent achievements can be found in [1, 2, 3] for classical CDMA. Regarding MT-CDMA the steady-state performance of MMSE (Minimum Mean Squared Error) linear and decision feedback (DF) joint detectors have been studied in [7, 8] respectively. In these papers, detectors including filters matched to both the symbol shape and the channel were assumed. The channel estimation was assumed to be ideal. In a practical system however channel estimation is an expensive device. Channel matched filtering and the associated channel estimation can be overcome if the forward section of the joint detector follows a symbol matched filter but works at an increased rate compared to the symbol rate. Hence a fractionally spaced device is obtained. In the present paper, we want to put some emphasis on fractionally spaced linear and decision feedback joint detectors (FS-DFJD) for multitone CDMA that have been presented in more details in [10, 11]. The design will be made for an MMSE criterion. The steady-state behavior will be analyzed and the performance of the detector will be assessed by means of the BER (bit error rate) for BPSK modulation. Asynchronous two-path channels will be considered.

2 System description

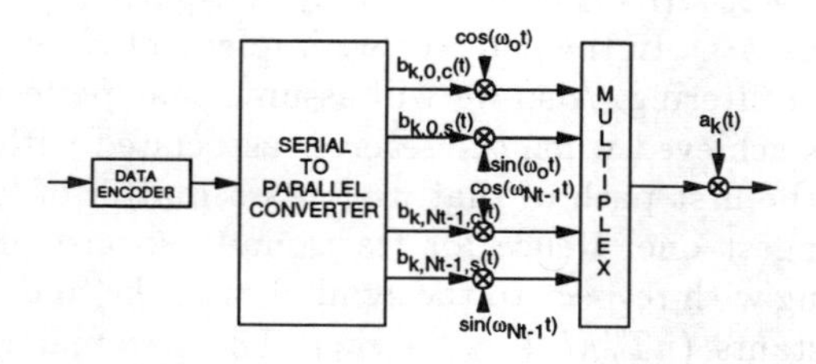

Fig. 1. Multitone CDMA Transmitter

The transmitter of a multitone CDMA is depicted in figure 1. It can be seen that the transmitted signal associated with a particular user is obtained by the modulation of N_t tones in parallel first and then spreading of the multitone signal by a user specific sequence having N_c chips per multitone symbol duration. In order to keep the bandwidth constant, the chip duration has to be constant and hence the ratio between the number of tones and the number of chips per symbol has to be constant. The lowpass equivalent signal transmitted by user k is given by

$$x_k(t) = \sqrt{\frac{2P}{N_t}} \sum_{p=0}^{N_t-1} \sum_{n=-\infty}^{\infty} I_{k,p}^{n}\, a_k(t)\, u(t-nT)\, \exp^{2\pi jpt/T} \tag{1}$$

where $a_k(t)$ is the waveform associated with the periodical pseudo-noise sequence multiplying the multitone signal and associated with user k. $I_{k,p}^{n}$ is the nth symbol conveyed by carrier p of user k and $u(t)$ is the symbol shape. It is assumed to be rectangular in the present paper. The RF frequency associated with the pth carrier is given by $f_0 + p/T$ where f_0 is some base frequency and T is the symbol duration. There are N_t tones and the factor $1/N_t$ in the power P is introduced to keep the overall power constant. We analyze the uplink. Assuming that the channel between user k and the receiver is linear with equivalent lowpass impulse response $c_k(t)$, and provided perfect carrier phase recovery, the received lowpass signal is given by

$$r(t) = \sum_{k=0}^{N_u-1} \sqrt{\frac{2P}{N_t}} \sum_{p=0}^{N_t-1} \sum_{n=-\infty}^{\infty} I_{k,p}^{n}\, h_{k,p}(t-nT) + n(t) \tag{2}$$

where N_u is the number of users, $u_{k,p}(t) = a_k(t)\, u(t) \exp^{2\pi jpt/T}$, $h_{k,p}(t) = u_{k,p}(t) \otimes c_k(t)$, and $\otimes$ denotes convolution. In the present work, we investigate the performance of the system for channel impulse responses given by $c_k(t) = \beta_{k1}\delta(t-\tau_{k1})+\beta_{k2}\delta(t-\tau_{k2})$ that is to say two-path impulse responses. The AWGN is denoted by $n(t)$. In the detector we implement $N_u \times N_t$ filters performing symbol matched filtering. Also we will assume that perfect synchronization (timing recovery) is achieved when the receiver associated with a particular user synchronizes with the first path of that user. As a matter of fact, the first path is usually the strongest one. Hence for fractionally spaced processing with an M-fold oversampling with respect to the symbol rate, the matched filter outputs are sampled at instants $(nT/M + \tau_{k1} + \epsilon_{k1})$. The parameter ϵ_{k1} accounts for possible imperfections of timing recovery. Hence the samples processed by the FS device are obtained by

$$z_{k,p}^{n} = \frac{1}{T\sqrt{2P}} \int_{-\infty}^{\infty} r(t)\, u_{k,p}^{*}(t - nT/M - \tau_{k1} - \epsilon_{k1})\, \mathrm{d}t \tag{3}$$

It appears that the overall system between the symbol generation up to the sampling of the matched filter outputs can be modeled by a Multiple Input Multiple Output (MIMO) digital equivalent system. Next, it also appears that the samples at the outputs of the matched filters can be affected by ISI, ICI and MAI. This suggests that the Joint Detection structure also has to be of the MIMO type.

3 Fractionally spaced linear joint detection (FS-LIJD)

The FS-LIJD investigated here builds an estimate $\hat{I}_{l,q}^{m}$ of a symbol associated with a certain tone of a certain user from the z samples associated with all tones of all users. Hence,

$$\hat{I}_{l,q}^{m} = \sum_{k=0}^{N_u-1} \sum_{p=0}^{N_t-1} \sum_{n=-K_1}^{K_2} c_{l,k,q,p}^{n}\, z_{k,p}^{mM-n} \tag{4}$$

In the present paper we investigate the design of the FS-LIJD structure for an MMSE criterion. According to the results reported in [5], we use the orthogonality principle. We require that

$$\mathbf{E}[\epsilon_{l,q}^{s}(z_{k,p}^{sM-n})^{*}] = 0 \qquad \text{for } -K_1 \leq m \leq K_2 \tag{5}$$

for all combinations of k, l, p, q and $\epsilon_{l,q}^{m} = I_{l,q}^{m} - \hat{I}_{l,q}^{m}$. This equation leads to a set of $N_u^2 N_t^2(K_1 + K_2 + 1)$ Wiener-Hopf equations. This set of equations can be written in a matrix form that depends also on the E_b/N_0 ratio.

4 Fractionally spaced decision feedback joint detection (FS-DFJD)

The FS-DFJD builds an estimate $\hat{I}_{l,q}^{m}$ of a symbol associated with a certain tone of a certain user from z samples associated with all tones of all users and from previous decisions associated with all tones of all users. Hence,

$$\hat{I}_{l,q}^{m} = \sum_{k=0}^{N_u-1} \sum_{p=0}^{N_t-1} \sum_{n=-K_1}^{0} c_{l,k,q,p}^{n} \, z_{k,p}^{mM-n} + \sum_{k=0}^{N_u-1} \sum_{p=0}^{N_t-1} \sum_{n=1}^{K_2} d_{l,k,q,p}^{n} \, \tilde{I}_{k,p}^{m-n}$$
$$= \mathbf{c}^T \mathbf{z} + \mathbf{d}^T \tilde{\mathbf{i}} = \begin{bmatrix} \mathbf{c}^T & \mathbf{d}^T \end{bmatrix} \begin{bmatrix} \mathbf{y} \\ \tilde{\mathbf{i}} \end{bmatrix} = \mathbf{c}_{eq}^T \mathbf{z}_{eq} \tag{6}$$

where $\tilde{I}_{k,p}^{m-n}$ is the decision about symbol $I_{k,p}^{m-n}$ and ^{T} denotes transposition. For mathematical tractability it is assumed that the decisions are always correct, and hence $\tilde{I}_{k,p}^{m-n} = I_{k,p}^{m-n}$. It has to be noticed that the forward section works at a higher rate (M) than the feedback section. We investigate the design of the FS-DFJD structure for an MMSE (minimum mean square) criterion. Using the matrix notation, and forcing to 0 the derivative of the objective function with respect to $\mathbf{c}_{eq}^T$, it turns out that the coefficients of the detector have to fulfill the following Wiener-Hopf equation:

$$\mathbf{R}_{\mathbf{y}_{eq}} \mathbf{c}_{eq}^T = \mathbf{r}_{\mathbf{yI}} \tag{7}$$

where $\mathbf{R}_{\mathbf{y}_{eq}}$ stands for the autocovariance matrix of vector $\mathbf{y}_{eq}$ and $\mathbf{r}_{\mathbf{yI}}$ stands for the mutual covariance vector between $\mathbf{y}_{eq}$ and $I_{l,q}^{m}$. If the coefficients of the equalizer fulfil the MMSE condition, it can be shown that the MSE is given by

$$\mathbf{E}\left[|I_{l,q}^{m} - \hat{I}_{l,q}^{m}|^2 \right] = \sigma_I^2 - \mathbf{r}_{\mathbf{yI}}^T \mathbf{R}_{\mathbf{y}_{eq}}^{-1} \mathbf{r}_{\mathbf{yI}}^* \tag{8}$$

This MSE may still be seen as a function of the channel, and of the basis functions that are selected to transmit the different symbols. *Hence it may be expected that a further step in the optimization of the system is the minimization of the MSE with respect to the basis functions, or to the codes.* This way of doing is also valid if we use FS-LIJD instead of FS-DFJD, but the optimal solution will not be the same. In other words, knowing that joint detection will be performed with a certain filter length, a further optimization can be achieved by selecting the "most appropriate" basis functions for that structure of the joint detector. As a matter of fact, a difficult point is associated with the time varying nature of the real radio channel.

5 Bit error probability

The residual interference after equalization can be computed as a function of the input symbols and the noise. By means of an approach based on characteristic functions, the bit error rate for BPSK can be computed exactly, as explained in [7, 8].

6 Computational results

6.1 Performance of FS-LIJD

Figure 2 shows the performance obtained for $N_u = 3$, $N_c/N_t = 31/4$ and $K_i = M = 4$. The delays of the different paths and their amplitudes have been set as presented in table 1. Gold codes are used. The averaged BERs curves for the 3

	τ_{k1}	τ_{k2}	β_{k1}	β_{k2}
k=1	0.00	0.2	1.0	1.0
k=2	0.25	0.35	1.0	1.0
k=3	0.50	0.60	1.0	1.0

Table 1. Parameters used for the computational results. The delays are normalized with respect to the symbol duration when $N_t = 1$

users are shown in the figure. We also have $K_1 = K_2 = M$ so that the absolute span of the filter is roughly constant. First of all the performance of the matched filter detector is very poor. As a matter of fact the interference is very high as the second path has the same amplitude as that of the first path. The other curves show the effectiveness of the FS-LIJD.

6.2 Performance of FS-DFJD

The performance of the DF-FSJD device has first been investigated assuming perfect timing recovery. This means that the ϵ_{k1}s are first assumed to be 0. Figure 3 provides results for $N_c/N_t = 31/4$. The level of performance is illustrated.

6.3 Sensitivity to timing errors

The sensitivity of the detector to timing errors has also been investigated. For this investigation a scenario with $N_u = 3$ users has again been considered. There were two paths per user with the same parameters as those used above. The timing of users 2 and 3 was perfect (meaning $\epsilon_{k1} = 0$ in the equations). There was a timing error for the reference user (user 1) varying between 0 and $0.05T_1$ where T_1 is the symbol duration when 1 tone is used. Results are provided by figure 4 for $N_c/N_t = 31/4$ and (M, K_1)=$(8, 8)$ and $(16, 16)$. It turns out that the sensitivity after FS-DFJD is reduced a lot for the system and scenario used in this set-up. It should also be mentioned that we assume rectangular chip in our model and consequently, the bandwidth of the signal is not limited.

6.4 Sensitivity to the near-far effect

A last issue which has been addressed is the sensitivity of the detector to the near-far effect. A near-far situation occurs when the different users are received with very different powers. For instance, a user far away from the base station might be hidden by an interfering user much closer to the receiver. In order to combat these effects, power control mechanisms have been investigated. Another possibility is to design detectors which are as robust as possible against this type of situations. In order to investigate the resistance of the FS-DFJD device the following set up was considered. The path gains of the reference user (user 1) were kept constant while the gains of all paths (2) of all users changed from -10 to +10 dB with respect to the path amplitude of the reference user. Figure 5 shows the BER sensitivity when $N_c/N_t = 15/2$ and various associations (M, K_1), with $K_2 = 1$. Besides $E_b/N_0 = 7\ dB$. It turns out that the reference user becomes quite rapidly (beyond $M = 2$) insensitive to a near-far effect in the scenario considered here.

7 Conclusions

The present paper has been devoted to the review of fractionally spaced linear and decision feedback detectors for CDMA and MT CDMA signals. Fractionally spaced devices with a forward section working at a higher rate than the symbol rate make it possible to avoid the channel estimation required to achieve channel matched filtering. The derivations have been made for an MMSE criterion. The performance of the detector has been analyzed for asynchronous two-path channels in steady-state conditions. It has been shown that the detector is very effective. Besides it has been pointed out that a further level of optimization could be achieved if the system would be optimized from the point of view of codes and/or basis functions. The sensitivity of the detectors to timing errors has also been investigated. Besides it has been demonstrated that such devices are resistant against a near-far effect.

References

1. *IEEE Journal on Selected Areas in Communications*, vol. 12, No. 4, May 1994.

2. *IEEE Journal on Selected Areas in Communications*, vol. 12, No. 5, June 1994.
3. *European Transactions on Telecommunications*, vol. 6, No. 1, January-February 1995.
4. R. Lupas and S. Verdu, "Linear multiuser detectors for synchronous code-division multiple access channels", *IEEE Transactions on Information Theory*, vol. 35, No. 1, January 1989, pp. 123-136.
5. J. G. Proakis, "Digital Communications", New-York, McGraw-Hill, 1989.
6. L. Vandendorpe, "Multitone spread spectrum multiple-access communications system in a multipath Rician fading channel", *IEEE Transactions on Vehicular Technology*, vol. 44, No. 2, May 1995, pp. 327-337.
7. L. Vandendorpe and O. van de Wiel, "Performance analysis of linear joint equalization and interference cancellation for multitone CDMA", *Wireless Personal Communications*, Special Issue on interference cancellation for Personal, Indoor and Mobile Radio Communications, Vol. 3, No 1-2, 1996, pp. 17-36.
8. L. Vandendorpe and O. van de Wiel, "Decision feedback multiuser detection for multitone CDMA", *European Transactions on Telecommunications*, ETT, vol. 8, No. 1, January-February 1997, pp. 67-78.
9. N. Yee, J.-P. Linnartz and G. Fettweis, "Multi-carrier CDMA in indoor wireless radio networks", *Proc. PIMRC'93*, Yokohama, September 1993, pp. D.1.3.1-D.1.3.5.
10. O. van de Wiel and L. Vandendorpe, "Fractionally spaced linear joint detection for multitone CDMA systems", *IEEE International Communications Conference ICC'96*, Dallas, June 1996, pp. 602-606.
11. L. Vandendorpe and O. van de Wiel, "Fractionally spaced decision feedback joint detection for multitone CDMA systems", IEEE Fourth International Symposium on Spread Spectrum Techniques and Applications, Mainz, Germany, September 1996, pp. 1361-1365.

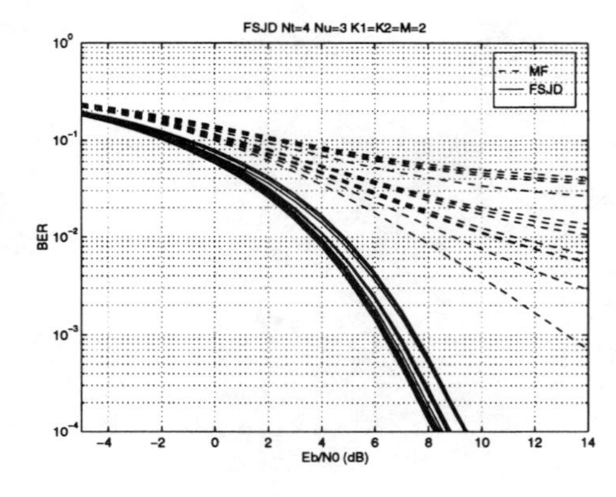

Fig. 2. BER obtained after FS-LIJD detection for $N_u = 3$, $N_c/N_t = 31/4$, $(M = K_i = 4)$

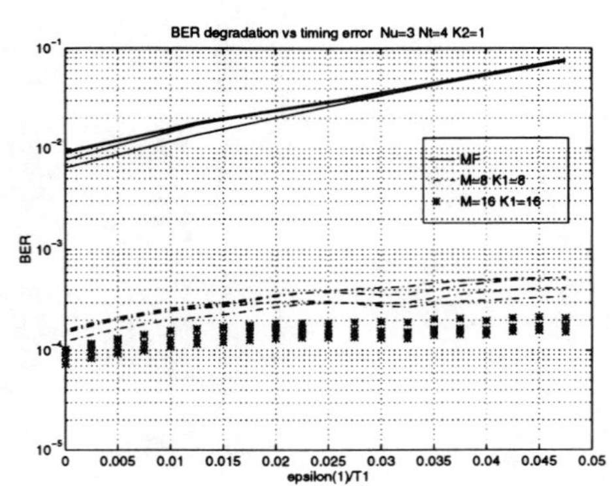

Fig. 4. BER degradation due to a timing error affecting a single user, for $N_u = 3$, $N_c/N_t = 31/4$, (M, K_1)=(8,8), (16,16) and $K_2 = 1$

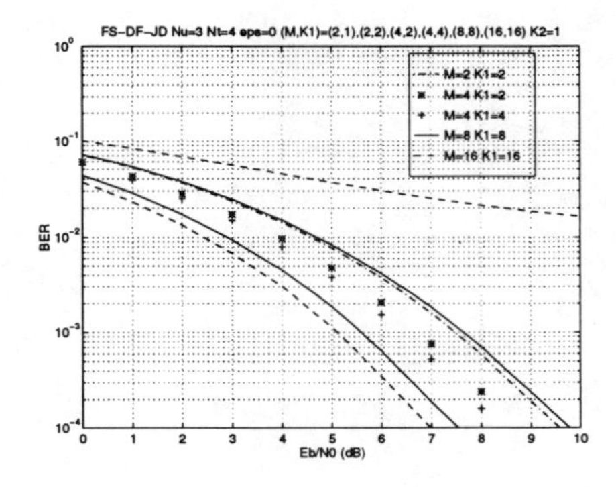

Fig. 3. BER obtained by means of FS-DFJD for $N_u = 3$, $N_c/N_t = 31/4$, $(M, K_1) = (2,1)$, $(2,2)$, $(4,2)$, $(4,4)$, $(8,8)$, $(16,16)$

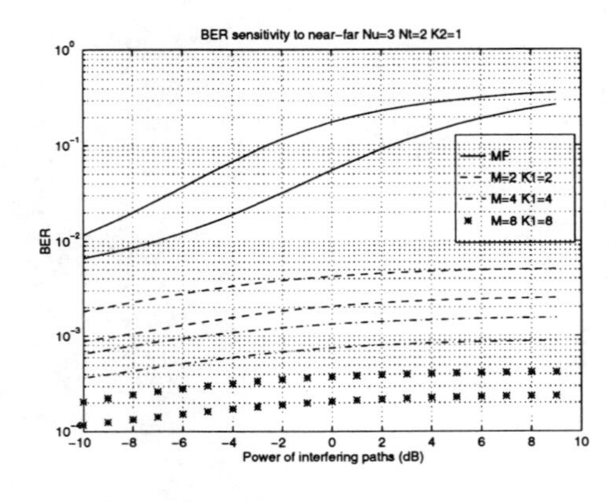

Fig. 5. BER evolution versus the near-far effect for $N_u = 3$, $N_c/N_t = 15/2$, $(M, K_1) = (2,2)$, $(4,4)$, $(8,8)$ and $K_2 = 1$

COMPARISON OF MULTIPLE ACCESS SCHEMES FOR AN OFDM DOWNLINK SYSTEM

Hermann Rohling, Karsten Brüninghaus, Rainer Grünheid

Technical University of Braunschweig
Schleinitzstraße 22, D-38092 Braunschweig, Germany
E-Mail: rohling@ifn.ing.tu-bs.de

ABSTRACT: Basic multiple access schemes as TDMA, FDMA and CDMA are considered for an OFDM (orthogonal frequency division multiplexing) downlink system. A comparison of these schemes is made in terms of flexibility, computation complexity, signalling overhead and bit error rate (BER).

INTRODUCTION

In the last years, the OFDM transmission technique has not only been applied to broadcast systems like digital terrestrial television (DTVB) and radio (DAB) but has also been discussed for use in mobile communications [1]. In this case, the access to the radio channel has to be organised and the question arises as to concept for an efficient multiple access scheme adapted to the OFDM technique.

From single carrier systems a variety of techniques is known, such as TDMA, FDMA and CDMA or combinations of them. Basically, these techniques can also be adapted to multicarrier transmission but some peculiarities have to be taken into account.

In this paper these three different multiple access methods are described and compared for the downlink of an OFDM communication system regarding flexibility, computation complexity, signalling overhead and BER.

K. Fazel and G.P. Fettweis (eds.), Multi-Carrier Spread-Spectrum, 23-30.

MULTIPLE ACCESS SCHEMES FOR OFDM

A schematic overview of multiple access schemes for OFDM is shown in Fig. 1. The allocation of the channel resources (time, bandwidth) to the individual users is visualized by different shades of gray. Every square in Fig. 1 represents the smallest resource unit to be allocated to a single user, i.e. one subcarrier for a length of a single OFDM symbol.

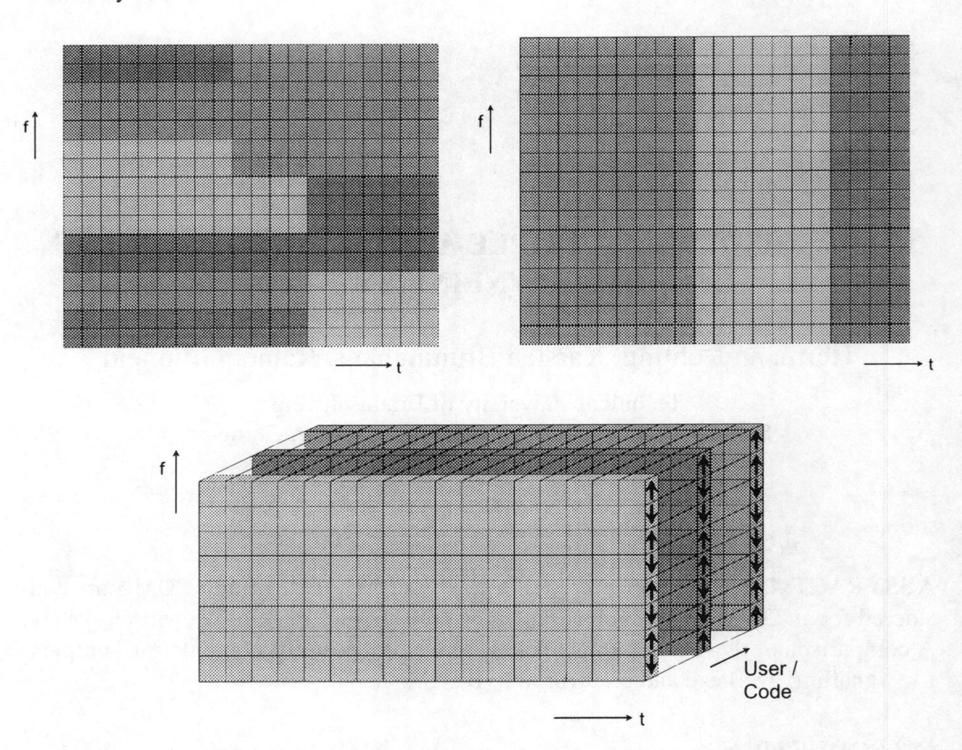

Figure 1: Overview of different multiple access schemes for OFDM: FDMA (top left), TDMA (top right), CDMA (bottom)

With OFDM-FDMA, see Fig. 1 top left, a single user allocates parts of the available bandwidth (one or more subcarriers) for at least one time slot, i.e. one OFDM symbol. The subcarriers occupied by a single user can be arbitrarily distributed within the available bandwidth. Since the total bandwidth is processed in an OFDM receiver a purely digital subcarrier management with high flexibility can be used, as it is discussed in the following sections. This is in contrast to a conventional single carrier system where generally a single *continuous* frequency band is allocated and processed.

In an OFDM-TDMA system, see Fig. 1 top right, the total bandwidth (all subcar-

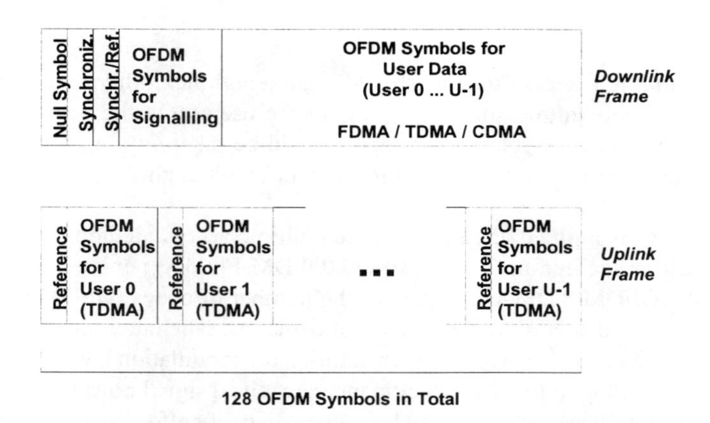

Figure 2: Frame structure of the considered OFDM-system

riers) is allocated to a single user for the duration of at least one OFDM symbol. The performance of an OFDM-TDMA system has been analysed in detail in [2].

Finally, in the case of OFDM-CDMA [3], see Fig. 1 bottom, which is also addressed as multicarrier CDMA, several users allocate the same bandwidth (subcarriers) using different orthogonal codes. Spreading is performed on data symbol level prior to IFFT-processing, therefore the OFDM-CDMA signal is basically an OFDM signal. To reduce the complexity of equalization in the receiver, M codesets of length $L = K/M$ (K = number of subcarriers) are usually used instead of a single codeset of length K.

COMPARISON OF MULTIPLE ACCESS SCHEMES

The comparison of the different multiple access schemes is based on the frame structure depicted in Fig. 2. A frame length of $F = 128$ OFDM symbols is assumed, applying a TDD structure with variable down-/uplink capacity per frame. TDMA is used in the uplink since FDMA and CDMA require high accurate frequency and block synchronization of all users which might be difficult to achieve in a cellular system. Furtheron, a number of $K = 256$ subcarriers is assumed spaced by $\Delta F = 28$ kHz. To cope with multipath propagation, a guard interval of length $T_G = 5\mu$s is added, yielding a total symbol duration of $T_S = 40.7\mu$s. In case of OFDM-FDMA and OFDM-TDMA, the subcarriers are DQPSK modulated and incoherently demodulated, whereas in the case of OFDM-CDMA, QPSK is applied. The total data rate resulting from these parameters amounts to 12.57 Mbit/s.

FLEXIBILITIY

An efficient multiple access scheme should grant a high flexibility in distributing the available time-bandwidth resources among all the users. On the one hand, the behaviour of the frequency-selective channel should be taken into account and on the other hand, the user requirements for different and/or changing data rates have to be met.

Concerning changing data rates all three multiple access methods show the same flexibility. Either the number of subcarriers (OFDM-FDMA) per user, or the number of time slots (OFDM-TDMA) per user or the number of codes (OFDM-CDMA) per user can be adapted according to the current demands. The latter technique assumes that all codes are known to every user. In addition the modulation level can be adapted in any case according to the user specific mean received signal power.

Since OFDM-CDMA exploits the frequency diversity effect by using several subcarriers for each data symbol no further subcarrier specific adaptation to the channel transfer function is possible and necessary. This is different with OFDM-FDMA and OFDM-TDMA. If the amplitude channel transfer function to every user is known in the base station (requiring a TDD system with OFDM-TDMA in the uplink), subcarrier adaptive modulation can be used in both cases, as analysed in [2]. OFDM-FDMA can additionally exploit the fact that each user experiences a different radio channel by allocating only "good" subcarriers with high signal-to-noise ratio to each user, thereby increasing the channel capacity.

COMPUTATION COMPLEXITY

The computation complexity which is associated with the different multiple access schemes mainly depends on the complexity of channel estimation and equalization. With OFDM-TDMA and OFDM-FDMA differential encoding can be applied requiring no equalization and channel estimation at all in the receiver if the signal is demodulated incoherently, thus leading to a very simple receiver structure. If no differential encoding is used in the transmitter channel estimation and equalization has to be performed in the receiver. Equalization is quite simple because only the phase shift and attenuation on every subcarrier has to be equalized (one tap equalizer) but channel estimation is relatively complex and requires additional pilot symbols to be embedded in the data stream. A comparison of both demodulation methods can be found in [4] showing that the performance gain of coherent demodulation is relatively low if the pilot symbol redundancy is used for channel coding in case of incoherent demodulation.

OFDM-CDMA as proposed in [3] requires coherent demodulation. Depending on the type of equalizer/detection algorithm the receiver can become very complex, especially if joint detection algorithms are used. The complexity of the simplest equalizer, i.e. a linear equalizer [5], is comparable with the one required for coherent OFDM-TDMA or OFDM-FDMA. In addition, despreading (correlation with the user-specific code) has to be performed in the receiver.

SIGNALLING OVERHEAD

Apart from computation complexity the signalling complexity of the different multiple access schemes has to be considered. Generally, some sort of signalling information about the subcarrier/code allocation has to be created at the base station and to be processed at the mobile receivers. If the resource allocation is managed by the base station (which is most effective), information about the subcarrier/code allocation for the down- *and* uplink has to be transmitted. The amount of signalling information in each frame is depicted in Tab. 1.

Access scheme	kind of signalling information	overhead [bit]
FDMA	allocation table for all subcarriers	$K \cdot \log_2(U)$
TDMA	start/end of time slot	$2U \cdot \log_2(F)$
CDMA	start/end of code index	$2U \cdot \log_2(K)$

Table 1: Signalling information required for different multiple access schemes; U = number of users, K = number of subcarriers, F = number of OFDM symbols per frame (up- and downlink)

The relative amount of signalling overhead (relative to the frame length of 128 symbols) using TDMA, FDMA or CDMA in the downlink and TDMA in the uplink is shown in Tab. 2 for different number of users. It is assumed that a code rate $R = 1/4$ is used for the signalling bits.

Access scheme	8 users	16 users	32 users
OFDM-FDMA	5.4%	7.7%	10.5%
OFDM-TDMA	1.4%	2.8%	5.4%
OFDM-CDMA	1.5%	3.0%	5.8%

Table 2: Relative amount of signalling overhead for different multiple access schemes in the downlink; $K = 256$ subcarriers, framelength $F = 128$ symbols, TDMA in uplink, code rate $R = 1/4$

Thus, it can be seen that OFDM-FDMA requires the largest overhead because a user number ($log_2(U)$ bits) has to be transmitted for each subcarrier. This overhead can be reduced if neighbouring subcarriers are clustered, however at the cost of reducing the effectiveness of adaptive subcarrier allocation.

BER PERFORMANCE

The BER performance of OFDM-TDMA, OFDM-CDMA and OFDM-FDMA was analysed on the basis of computer simulations using punctured convolutional codes with soft decision decoding.

The radio channel was modelled to be frequency-selective and time variant. The power delay profile is exponentially decreasing with a maximum delay of $\tau_{\max} = 5\mu s$. For the Dopppler spectrum a Jakes distribution with a maximum doppler frequency $f_D = 70\text{Hz}$ was chosen.

A number of U=8 users is analysed as an example, where each user transmits the same data rate in the downlink. For OFDM-TDMA, this implies an unique number of OFDM symbols in each downlink frame (all subcarriers used).

In case of OFDM-FDMA, the total bandwidth is divided into 8 groups of 32 subcarriers. It is assumed that the base station is able to measure the attenuations $|h_i^u|$ of each subcarrier of user u during the uplink of this user. A realistic subcarrier allocation is used according to the following algorithm: In the base station the subcarriers of each user are first sorted according to their attenuation serving as a "favourite list". To maximize the transmitted signal power, each user tries to allocate as much as possible of the (in this case 32) subcarriers standing on top of his favourite list. The series of service depends on the ratio of allocated to requested number of subcarriers, i.e. the user with the smallest ratio has the right to allocate one new subcarrier (going down his list from top to bottom). In case this subcarrier is already occupied by another user, it is finally allocated to the one who would otherwise suffer from the largest accumulated relative power loss P_a^u, defined as

$$P_a^u = \frac{|h_{l(i)}^u|^2 - |h_{l(j)}^u|^2 + h_{acc}^u}{\sum_{i=1}^{K/U} |h_{l(i)}^u|^2}$$

Here, the index $l(i)$ denotes the index of the currently wanted subcarrier (i-th element of favourite list) of user u while the index $l(j)$, $j > 32$, denotes the index of best alternative subcarrier (out of the bottom 224 subcarriers of the list) which is not yet allocated. h_{acc}^u is the accumulated absolute power loss. Using the relative power loss as allocation criterion assures that not all of the best 32 subcarriers of one user can be occupied by other users as it could happen if the overall transmitted signal power is maximized. In the simulations an ideal knowledge of the amplitude channel transfer factors is assumed for every link.

As for OFDM-CDMA, each user is assigned 32 Walsh-Hadamard codes. The code length L is set to 16 (with full load, i.e. 16 codes per group). Wiener filtering (see [5]) is used as a detection method. When coherent demodulation is applied, ideal knowledge of the channel transfer function is assumed.

In Fig. 3, 4 and 5, the performance of OFDM-TDMA, -CDMA, and -FDMA is plotted when using convolutional codes (memory length 6) with different code rates. It can clearly be seen that OFDM-CDMA leads to a better performance than OFDM-TDMA, due to different demodulation schemes and an additional diversity

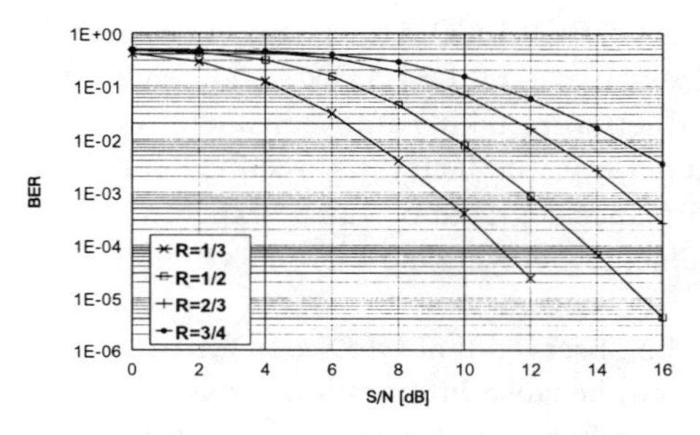

Figure 3: Performance of OFDM-TDMA; incoherent demodulation

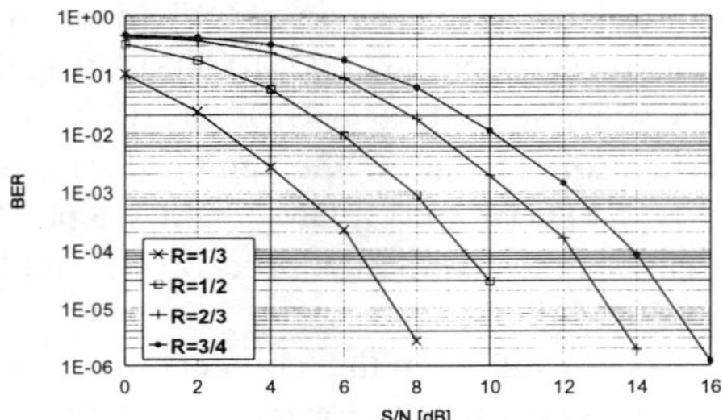

Figure 4: Performance of OFDM-CDMA; coherent demodulation, all codes used

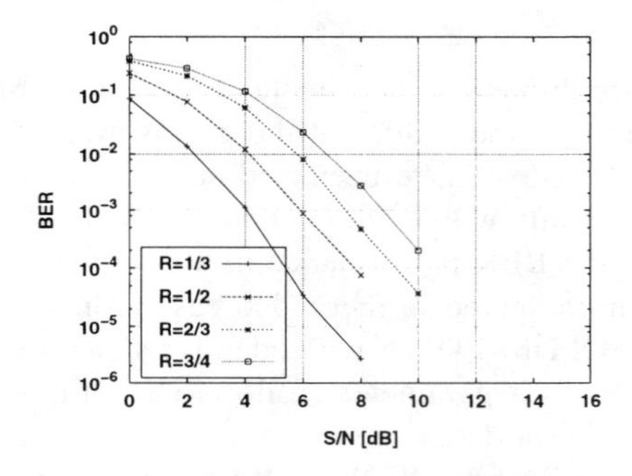

Figure 5: Performance of OFDM-FDMA; incoherent demodulation, realistic subcarrier allocation

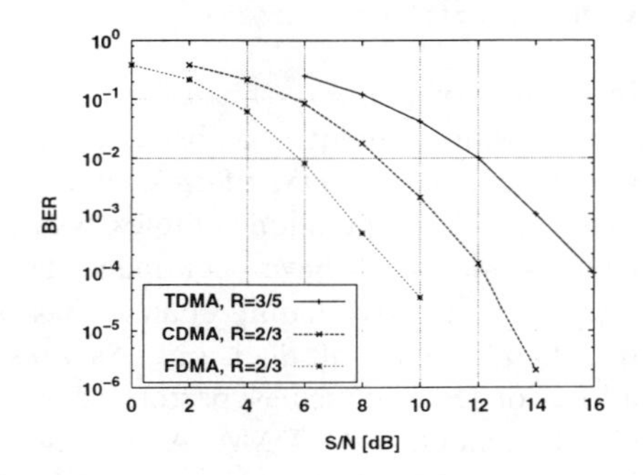

Figure 6: Performance comparison of OFDM-TDMA, -FDMA and -CDMA for a fixed user data rate

gain for CDMA. A difference of about 4 dB is observed at a BER of 10^{-3}. From Fig. 5 it becomes obvious that FDMA outperforms both CDMA and TDMA. In a realistic system a performance loss for both OFDM-FDMA and OFDM-CDMA has to be taken into account due to imperfect knowledge of the (amplitude) channel transfer factors.

A coherent demodulation in the case of OFDM-CDMA implies that pilot symbols are inserted in an OFDM symbol. This redundancy can, on the one hand, be used for a more powerful channel coding in the case of incoherent demodulation as mentioned before because no channel estimation is required here. On the other hand OFDM-FDMA requires considerably more signalling overhead than OFDM-CDMA and OFDM-TDMA if the subcarriers are allocated separately to the users.

Thus, for the considered frame structure a comparison at the same net data rate leads to a comparable code rate for FDMA (incoherent) and CDMA (coherent) if $\approx 7\%$ pilot symbols are used for channel estimation in case of CDMA. Since TDMA (incoherent) requires neither pilot symbols nor a large signalling overhead a code rate $R = 3/5$ can be used instead. The corresponding results for a fixed user data rate are summarized in Fig. 6.

The performance gain of OFDM-FDMA vs. OFDM-CDMA amounts to 3 dB at a BER of 10^{-4}. While OFDM-CDMA is still superior to OFDM-TDMA, the remaining difference is 4 dB at a BER of 10^{-4}. Furthermore, taking into consideration that perfect knowledge of the (amplitude) channel transfer factors is assumed for both OFDM-FDMA and OFDM-CDMA, a performance degradation can be expected when a real channel estimation procedure is implemented shifting the BER curves in favour of OFDM-TDMA.

As a result from this comparison, it can be observed that OFDM-FDMA leads to the best performance for the considered system because information about the radio channel is taken into account, allocating only suitable subcarriers with high SNR to each user.

CONCLUSIONS

In this paper, the performance of different multiple access schemes for the OFDM transmission technique has been studied. The characteristics of the access methods OFDM-TDMA, OFDM-FDMA, and OFDM-CDMA have been outlined. Some remarks about computation complexity and flexibility with regard to data rate and radio channel adaptation have been made, before the BER performance in a time-variant frequency-selective fading channel has been compared in the coded case (convolutional codes with soft decision). As a result, OFDM-FDMA with adaptive subcarrier allocation leads to the best performance for the given parameters, followed by OFDM-CDMA and OFDM-TDMA. Assuming a fixed user data rate, a gain of about 3 dB of FDMA vs. CDMA at a BER of 10^{-4} is observed, while OFDM-TDMA performs approx. 4 dB worse. If a loss due to real channel estimation is considered, the difference to OFDM-CDMA and OFDM-FDMA can be expected to shift in favour of OFDM-TDMA. Finally, it should be emphasized that the analysed TDMA and FDMA approach with incoherent demodulation can do without a channel estimation procedure, thus reducing the computation complexity in each mobile receiver.

References

[1] C.Reiners, H. Rohling: " Multicarrier Transmission Technique in Cellular Mobile Communication Systems". Proc. IEEE VTC'94, pp. 1645-1649, Stockholm, 1994

[2] H. Rohling, R. Grünheid: "Performance of an OFDM- TDMA Mobile Communication System", Proc. VTC 1996, Atlanta, pp. 1589-1593

[3] N. Yee, J.-P. Linnartz, G. Fettweis: "Multicarrier-CDMA in Indoor Wireless Radio Networks". PIMRC'93, Yokohama, pp. 109-113, 1993

[4] H. Rohling, T. May: "Comparison of PSK and DPSK modulation in a coded OFDM-system". To appear in Proc VTC 1997

[5] N. Yee, J.-P. Linnartz: "Wiener Filtering of Multi-Carrier CDMA in Rayleigh Fading Channel", Proc. PIMRC 1994, The Hague, pp. 1344-1347

LINK/SYSTEM LEVEL PERFORMANCE OF AN OFDM/CDMA DOWNLINK CONFIGURATION

J.P. Castro[1], A. Toskala[2], L. Chalard[1], S. Hämäläinen[2], and K. Kalliojärvi[2]

[2]Centre Suisse d'Électronique et de Microtechnique, CSEM, Jaquet-Droz 1, CH-2007 Neuchâtel, Switzerland
[2]Nokia Research Center, P.O.Box 407, FIN-00045, NOKIA GROUP, Finland

ABSTRACT

In this paper the link and system level performance of a cellular OFDM/CDMA downlink is evaluated. A system level simulator is built to utilise the link level simulation results and to derive the actual network capacity. Three service transmission rates, i.e., 12 kbits/s, 144 kbits/s and 2 Mbits/s have been studied in the radio link level for the evaluation of the suitability of OFDM/CDMA as an UMTS multiple access method. In the Link side the configuration of the sub-carrier allocation does influence the BER performance. From the system level results, it can be concluded that for cellular applications OFDM/CDMA exploits better reuse values higher than 1 as well as dynamic channel allocation options, when the power control mechanism is not optimised.

INTRODUCTION

The requirements for the Universal Mobile Telecommunications System (UMTS) capability set high demands on the air interface of the 3rd generation mobile networks. In Europe, within the Advanced Communications Technologies and

K. Fazel and G.P. Fettweis (eds.), Multi-Carrier Spread-Spectrum, 31-38.

Services (ACTS) programme, the FRAMES (Future Radio Wideband Multiple Access Systems) project has the objective to define and specify a UMTS air interface [1]. The OFDM/CDMA model has been studied as one of the downlink options in the FRAMES Multiple Access (FMA) system [2].

OFDM has gained a lot of interest in the field of communications, where it has been used for example in broadcast applications like DAB or various DTV solutions [4]. OFDM has also been considered earlier for 3rd generation mobile communication systems [7]. The methods of OFDM studied in this paper follows the latter applications using OFDM/CDMA modes to combat frequency selective multipath fading, and to offer more resistance towards co-channel interference typical to cellular environments. The model utilises MLSE type detector, although other detector types including interference cancellation used in [3] can also be applied.

To calculate the capacity of the multiple access option, which is a key criteria in the air interface evaluation process, link level performance parameters such as E_b/N_o for given BERs, are used. These types of link level measurements for different schemes have solid ground because well tested signal processing simulation programs can be used to obtain results of various multiple access techniques. However, the system level issues are more complicated, because each multiple access scheme has its own demands on the modelling to provide capacity calculations. Thus, dedicated tools or programs are often designed to adequately estimate capacity, and to obtain other useful information on the strengths and weaknesses of a particular multiple access scheme. For the OFDM/CDMA downlink model this is precisely the case, special adaptations were made to a system level program to access its performance. Therefore, in the following we will briefly present the link and system level methods utilised in the evaluation process.

THE LINK LEVEL CONFIGURATION

Due to the coherence time of the channel, we configure the OFDM/CDMA model symbols to have a duration of 409.6μs. Using a bandwidth of 2.5MHz, this leads to 1024 carriers with 2.44kHz spacing. This number of carriers allows the use of the FFT/IFFT algorithm for the OFDM demodulation/modulation.

formed with 1024 subcarriers, using 2.5 MHz bandwidth resulting in 409.6μs duration. Guard intervals of 4 μs were inserted between successive OFDM/CDMA symbols to avoid intersymbol interference from the channel multipath. The transmitted signal on N subcarriers of the qth OFDM symbol can be given as

$$y_q = C s_q = [c_1 c_2 \ldots c_u] s_q. \qquad (1)$$

where the N x u matrix C is the spreading code matrix, consisting of u length N spreading sequences as its columns. s_q is an u vector consisting of the symbols transmitted with these u different spreading codes. The introduction of CDMA component brings frequency diversity to OFDM and allows to better combat the frequency selective fading on the subcarriers. With long enough guard interval and the OFDM symbol being shorter that the time-coherence of the channel, the fading on each subcarrier will appear flat fading type and can be then modelled with a complex attenuation term. Thus the received and de-multiplexed signal is given as

$$z_q = A_q C_q s_q + w_q. \qquad (2)$$

where $A_q = diag(a_q,1 a_q,2 \ldots a_q,N)$ is the channel co-efficient matrix, where the coefficients are time varying complex numbers.

In the evaluation process, the subcarriers were grouped in 256 sets of 4 subcarriers, each of them conveying 8 bits with specific codes to provide frequency diversity. Figure 1 illustrates the frequency allocation of the sets, where each spread block is assigned to one specific set. It also illustrates other configuration parameters.

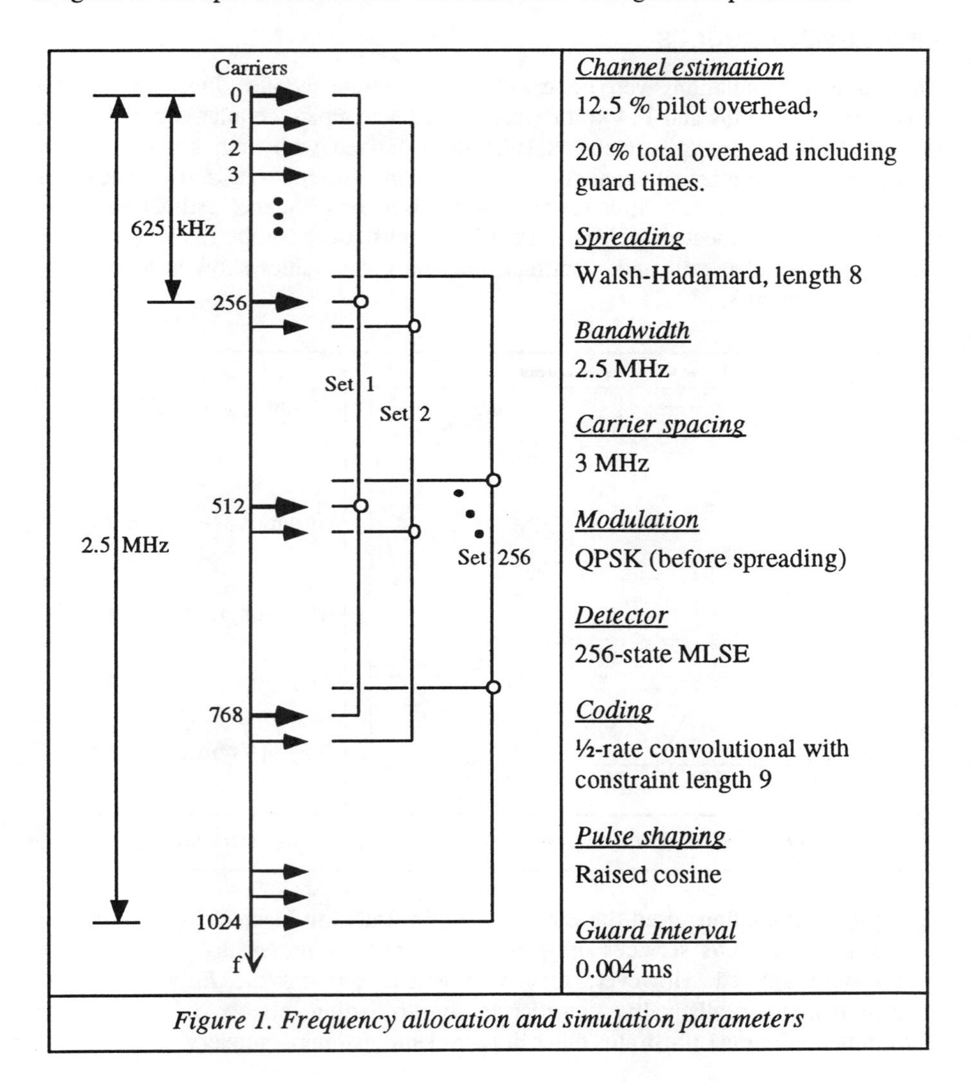

Figure 1. Frequency allocation and simulation parameters

The Maximum Likelihood Sequence Estimator (MLSE) method evaluates, for each set of 4 subcarriers, the most likely transmitted sequence s_q by minimising the squared Euclidean distance d^2 between the received and all possible transmitted sequences. Other sub-optimal and simpler approaches can also be found in [6].

The motivation to apply OFDM/CDMA techniques in this paper only to the downlink was the facility of implementation, and the offered flexibility in terms of service mapping. Also low level of intra-cell interference was ensured. Furthermore, services mapping benefit from maximum diversity, by means of hopping, interleaving and distribution of the data on several subcarrier sets. In the uplink especially the synchronisation is more demanding.

LINK LEVEL RESULTS

The link level simulations were done with three different services. In the Macro-cell environment 12 kbit/s and 144 kbit/s were studied, and a pico-cellular channel model was used for 2 Mbits/s services. Results were derived with both known channel parameters and with actual channel estimation having different ratios of overhead for pilot subcarriers. Figure 2 illustrates non-optimised real channel estimation E_b/N_o values next to the ideal or perfect estimation results. Because the real estimator of the present model is still under optimisation, the given values show a degradation higher than two dBs.

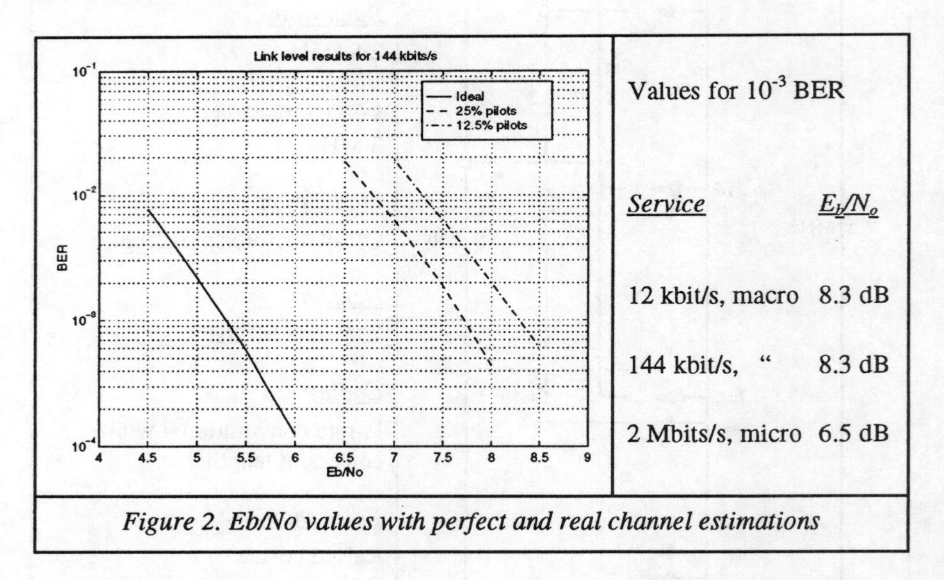

Figure 2. Eb/No values with perfect and real channel estimations

The system simulations used the E_b/N_o values illustrated in Figure 2. Notice that for the 12 and 144 kbit/s services the real channel non-optimised channel estimation values were applied. However, for the 2 Mbit/s services the E_b/N_o level was estimated more realistically considering lower fading effects in the indoor environment. Figure 1 illustrates other link level simulation parameters.

In the preceding link level results each service was specified by its user data rate and its maximum end-to-end delay. The latter took into account the coding time, interleaving of the corresponding symbols, and the complementary processes at the receiver. 12 kbit/s with 40 ms delay represented telephone or voice services; while 144 kbit/s with 100 ms delay represented data services. High data rate transmission at 2Mbits/s did not have delay restrictions; however, it did not exceed 40 ms. Interleaving had higher impact on E_b/N_o values of the lower rate services than those of the 2 Mbits/s transmissions. Finally, it should be mentioned that the link level results presented in this paper are initial values with models simulated at full loads.

SYSTEM LEVEL SIMULATOR

The system level performance for macro cell environment was evaluated following the CDMA approach, and using "Dynamic Channel Allocation" concepts. Micro and pico/indoor cellular environments at the system level are not covered in this paper.

The following pathloss notation for the macro environment was used, with the assumption for the carrier frequency to be around 2.0 GHz.

$$L = 29 + 36\log_{10}(R) + 31\log_{10}(f) \tag{3}$$

The basic equation for the C/I of an user and a carrier with OFDM/CDMA in the case of synchronously arriving signals is:

$$(C/I)_i = \frac{P_{Rx,i} G_p}{\sum_{\substack{j=0 \\ j\neq i}}^{N} \alpha_j \cdot P_{Rx,j} + \sum_{k=0}^{M} \beta_{IC} \cdot P_{Rx,k}^{tot} + N_0}. \tag{4}$$

In the above equation, $P_{Rx,i}$ is the receiver power of the carrier i. $P_{Rx,k}^{tot}$ is the total received power from BTS k. G_p is the processing gain, α_j is the orthogonality factor for intra-cell interference, β_{IC} models the orthogonality loss due to un-ideal channel estimation and due to fading multipath channel, and N_o models the thermal noise. Since the system level simulations model the interference-limited case, N_o can be neglected. The orthogonality factor α_j models how much the jth subcarrier that was received from the same BTS interferes the observed carrier.

We can assume orthogonality inside the cell, since intra-cell interference was already included in the E_b/N_o values from the link level simulations. Thus, the term α_j including its own term interference, i.e. the first two terms in the denominator of equation (4), will be equal to zero. Then the *(C/I)$_i$* can be expressed as

$$(C/I)_i = \frac{P_{Rx,i} \cdot G_p}{\left[\sum_{k=0}^{M} \beta_{IC} \cdot P_{Rx,k}^{tot}\right] \cdot \gamma}. \tag{5}$$

The parameter γ in equation (5) models the orthogonality between the signals from different BTSs. For inter-cell interference, DS-CDMA type interference was assumed, where the interference spreads over the whole bandwidth, because signals come via different channels with different propagation delays and multipaths and do not remain orthogonal considering that the subcarrier set allocation was random. In addition, because the network is also asynchronous, both β_{IC} and γ will have value 1. However, since the interference is not always spread over the whole bandwidth the value for γ would then be less than one and the bandwidth considered for the interference smaller than the total bandwidth. The equation for a single becomes

$$C/I = \frac{P_{Rx}G_p}{\left[\sum_{k=0}^{M} \beta_{IC} \cdot P_{Rx,k}^{tot}\right] \cdot \gamma}, \tag{6}$$

where

$$P_{Rx} = \sum_{i=0}^{N} \frac{P_{Rx,i}}{N}. \tag{7}$$

The corresponding Eb/No expression to equation (6) can be defined as

$$E_b / N_o = \frac{P_{Rx}G_p}{\left[\sum_{k=0}^{M} \beta_{IC} P_{Rx}^{tot}\right]\gamma}. \tag{8}$$

The power control concept is such that all the subcarriers (all the users) have the same transmission power within a BTS. The transmission power used depends on the measurements performed by all the mobiles. An approach where parts of the OFDM/CDMA symbols have slightly higher power, was also tested, but resulted in lower orthogonality inside the OFDM/CDMA symbol in the link level.

Thus, in the power control algorithm the C/I was measured for each mobile and the power balancing was made without command errors. The power control error due to the errors in measurements, channel state changes and errors in signalling, was modelled by adding a lognormal variable with 3 dB variance to the transmission power level. In the macro-cellular environment, the following pathloss notation was used, with the assumption for the carrier frequency to be around 2.0 GHz.

The shadow fading in the macro-cellular environment was modelled as a lognormally distributed variable with standard deviation 10 dB. The macro-cellular base stations were placed on a hexagonal grid.

We achieve "*Dynamic channel allocation*" in a particular manner, namely by taking advantage of the frequency reuse on the cells (e.g. a three colour clustering on the hexagonal cell geometry). Furthermore, inside each frequency reuse layer, we considered that channel- (or subcarrier) assignment was done in such a way that no intracell interference occured.

Thus, the interference was modelled assuming that OFDM on many carriers over a large bandwidth (i.e. large with respect to the data rate) can be modelled as a CDMA system with a spreading factor given by (*Total Subcarriers / Subcarriers per user*). For example, given the link level bandwidth of 2.5 MHz and 1024 subcarriers, the bandwidth per subcarrier results in 2.44 kHz. Using QPSK modulation, ½-rate coding and a bit rate of 12 kbit/s (voice), this leads to about 4.91 subcarriers per user. This leads to (1024/4.91) = 208.3 . This means that interference between two users will be relatively small, because of the separation of subcarriers, specially if we take into account the spreading and the channel configuration illustrated in Figure 1.

CAPACITY RESULTS

Table 1 illustrates the values from the cellular capacity simulations for 12 kbits/s, 144 kbits/s with and without Dynamic Channel Allocation (DCA) techniques.

Table 1 Spectral efficiency values for 5 % outage

Reuse Factor	Service	Spectral efficiency[kbit/s/cell/MHz] no DCA	with DCA
1	12 kbit/s	58 (macro)	
1	144 kbit/s	60 (macro)	
3	12 kbit/s		88 (macro)
4	12 kbit/s		105 (macro)
4	144 kbit/s		86.4 (macro)
7	144 kbit/s		174.85 (macro)

Clearly the results illustrated in Table 1 when including the DCA concept show improvement over the classical approach. In the spectral efficiency values, scaling corresponding to 3 MHz carrier spacing was assumed. The "-3dB" bandwidth is 2.75 MHz, but the actual carrier spacing to be used depends on the requirements set to the spectral mask and also from the spectral spreading due to the power amplifier. In the base station transmitter the use of highly linear amplifier is not considered to be that critical when compared to the mobile terminal and thus moderate spectral spreading can be expected and the required carrier spacing is supposed to be 3 MHz or less.

Based on the studies it could be noted that reuse 1 with OFDM/CDMA does not provide optimum results, therefore a higher factor is clearly needed for efficient operation. In the above results, reuse factors of 3 and higher have been used.

The power control algorithm used had problems since the users share the same OFDM/CDMA symbol. Thus, setting different power level for different users was difficult. This resulted in that for part of the users the received signal level was too high, and thus generating unnecessary interference to the surrounding cells.

CONCLUSIONS

The link and system level performance of cellular OFDM/CDMA downlink was evaluated in this paper. As a multiple access method the OFDM/CDMA can fulfil UMTS requirements in providing the required range of services up to 2 Mbits/s.

From the system level results, it can be concluded that for cellular applications OFDM/CDMA requires reuse factors higher than 1, and should benefit from dynamic channel allocation, when the power control mechanism is not optimised. Also, having several users sharing the same OFDM/CDMA symbol in environments with low cell isolation causes degradation in capacity, since the power control implementation for several low rate users is difficult. The power control commands needed when coupled with DS-CDMA in the uplink caused some limitations since the OFDM/CDMA symbols used in the present model were too long for the fast power control mechanisms used. Thus command rates in the order of 1 or 2 kHz could not be practically implemented in the downlink transmission for the mobile direction. However, it should be noted that the symbol length could be reduced in size with other configurations. Therefore, when using adaptive power control and shorter symbols, OFDM/CDMA downlink coupled with DS-CDMA using multi-user detection techniques in the uplink, would be an interesting alternative to balance restrictions seen in typical DS-CDMA systems. For example, OFDM/CDMA based systems tend to be less susceptible to traffic loads.

ACKNOWLEDGEMENT

This work has been partially funded by the European Community in the ACTS program under the ACTS AC090 FRAMES project. The authors acknowledge the contributions of colleagues from Siemens AG, Roke Manor Research Limited, Ericsson Radio Systems AB, Nokia Corporation, Technical University of Delft, University of Oulu, France Telecom CNET, Centre Suisse d'Electronique et de Microtechnique SA, Swiss Federal Institute of Technology Zurich, University of Kaiserslautern, Chalmers University of Technology, The Royal Institute of Technology, Instituto Superior Técnico.

REFERENCES

[1] Ojanperä et al.,"FRAMES - Hybrid Multiple Access Technology", Proceedings of ISSSTA'96 Mainz, Germany. September 1996.

[2] T. Ojanperä et al., "Comparison of Multiple Access Schemes for UMTS", Proceedings of VTC'97, Phoenix, Arizona, USA, May 1997.

[3] K. Kalliojärvi, "Interference Cancellation in OFDM-CDMA Systems", Proceedings of IEEE Nordic Signal Processing Symposium, September 24-27.1996, Espoo, Finland.

[4] M. Alard and R. Lassalle, "Principles of Modulation and Channel Coding for Digital Broadcasting for Mobile Receivers," EBU Review, Tech. No. 224, Aug. 1987, pp. 47-69.

[5] K. Fazel, S. Kaiser, P. Robertson and M.J.Ruf, "A Concept of Digital Terrestrial Television Broadcasting," Wireless Personal Communications, vol. 2, No. 1 pp. 9-27, 1995.

[6] S. Kaiser, "On the Performance of Different Detection Techniques for OFDM-CDMA in Fading Channels," In Proceedings of Globecom'95, Singapore, pp. 2059-2063.

Section II

APPLICATIONS OF MULTI-CARRIER SPREAD-SPECTRUM

PERFORMANCE ANALYSIS OF A NEW MULTI-CODE AND MULTI-CARRIER HYBRID TRANSMISSION SCHEME FOR FUTURE BROADBAND MOBILE COMMUNICATION SYSTEMS

Hiroshi Harada and Ramjee Prasad

Telecommunications and Traffic Control Systems Group
Delft University of Technology
P.O.Box 5031, 2600GA Delft, The Netherlands
Tel./Fax: +31 15 278 2147/1774
E-Mail:H.Harada@et.tudelft.nl

Abstract- This paper proposed a new Multi-carrier and Multi-code hybrid transmission scheme that used cyclic code generated from a modified M-sequence to keep the orthogonality of codes not only between parallel channels but also between direct and delayed waves, and to reduce the number of kinds of codes used by a single user. The BER performance in Rayleigh fading envirnment is evaluated by computer simulations. Computer simulation results show that our proposed scheme achieves good BER performance in the fast selective Rayleigh fading environment and is suitable for future broadband land mobile communication systems.

1. Introduction

For broadband mobile communication of several Mbps, BER performance is severely affected by multi-path propagation characteristics because the multipath delay spread causes intersymbol interference (ISI). The ISI imposes an

K. Fazel and G.P. Fettweis (eds.), Multi-Carrier Spread-Spectrum, 41-48.

upper limit on the data transmission rate. One candidate to overcome multipath environment is to use parallel transmission, in which high-speed serial data is converted to slow parallel data in several channels to suppress the delay time within symbol period. These parallel data are multiplexed using Frequency Division Multiplexing (ex. Multi-carrier or OFDM transmission[1]) or Code Division Multiplexing (ex. Multi-code transmission[2][3]) or these hybrid transmission system[4]. Especially, recently there has been increasing research for hybrid transmission system, because of its high bandwidth efficiency.

To realize high quality hybrid transmission, we must summarize many problems included in both parallel transmission schemes. As for Multi-carrier (OFDM) transmission, the ISI can be eliminated by inserting a guard interval between successively transmitted signals, which makes a complicated equalization unnecessary. However this system must be equipped with guard interval insertion and removal circuits. In addition OFDM technique is very sensitive to the carrier frequency offset between the transmitter and receiver. The frequency offset destroys the orthogonality among the subcarriers. Therefore we must estimate both the amount of frequency offset and propagation characteristics.

On the other hand, as for Multi-code transmission, a number of studies have been made, because it is possible to use RAKE reception for the delayed wave, and it is easy to apply to CDMA. However, there are several problems that if conventional multi-code transmission is implemented to broadband mobile communication system. One of problems is that the orthogonality of codes between direct and delayed waves is not preserved, and the interference noise is generated from delayed waves even if the orthogonality of codes between parallel transmission channels is kept. In addition, many kinds of orthogonal codes are necessary for each user, although the number of orthogonal codes is limited. In this paper, by taking these problems includes in hybrid transmission into account, we propose a new multi-carrier and multi-code hybrid transmission scheme that used cyclic code generated from a modified M-sequence to keep the orthogonality of codes not only between parallel channels but also between direct and delayed waves, and to reduce the number of codes used by a single user.

This proposed system has several advantages. Firstly the guard interval has already inserted on the procedure of code generation and at the receiver we need only DFT circuit and code correlator. Therefore this system need not to equip the guard interval insertion and removal circuits. In addition, it is very easy for proposed transmission scheme to estimate delay characteristics, synchronization point of spread code and start point of DFT. Besides these advantage, we can realize high capacity transmission scheme. In this paper, moreover, computer simulations show that proposed scheme can estimate the delay profile exists within guard interval and these direct and delayed waves are combined by Rake receiver, therefore our scheme can achieve broadband transmission with high quality even in very fast Rayleigh fading channel.

2. Configuration of the proposed multi-code transmission scheme

2.1 Problems in conventional multi-code transmission scheme

Figure 1(a) shows the waves received at a receiver and the output of the correlator when using the parallel transmission scheme with conventional orthogonal codes. In this figure, it is assumed that the transmitted serial data is converted to parallel data in 2 channels, spread by two orthogonal codes $C_1(t)$ and $C_2(t)$, and multiplexed.

In a multipath fading environment, a direct wave and a delayed wave with delay time of T_D are received. If the received direct waves are correlated with the code $C_1(t)$, the correlator output of channel 2 does not interfere with that of channel 1 at the synchronization point of T_{syn} because the orthogonality is kept between these two codes. However, the interference noise is generated by the delayed waves because they are not orthogonal to the code $C_1(t)$ at the synchronization point of T_{syn} as shown in Fig.1(a). Therefore the BER performance gets worse. In addition, many kinds of codes are required to transmit the broadband data in this scheme although the number of orthogonal codes is limited.

2.2 Code configuration of the proposed mult-code transmission scheme

We are confronted by two problems. The first is that it seems very difficult to keep the orthogonality of codes between direct and delayed waves. The second is that we need a lot of kinds of codes as the number of parallel transmission channels increases.

To cope with the first problem, we propose to use a new code generated by giving a constant direct current bias to the M-sequence. Though the M-sequence has autocorrelation characteristics illustrated by the dashed line in Fig. 2, this biased M-sequence becomes as illustrated by solid line. The bias value, α, is given by

$$\alpha = \frac{1}{N}(-1 \pm (1+N)^{1/2}), \tag{1}$$

where N is the code length. Moreover the guard chips are added before and after this biased M-sequence. This guard chips are extended by repeating the same biased M-sequence. In this paper, this new sequence added to constant dc bias and guard chips is named as 'modified M-sequence'[5]. At the receiver, the spread signal is correlated with the biased M-sequence without the guard chips. Figure 2 shows an example of the case of 2 chips as guard chip.

There remains a second problem: we need many kinds of codes. To cope with this problem, we use cyclic codes of a modified M-sequence for the other channels, where the cyclic shift interval is equal to the number of guard chips plus one chip. This system has the advantage as follows: there is no interference between parallel transmission channels if delay time doesn't exceed guard chips. Figure 3 shows an example of cyclic modified M-sequence code with 2 guard chips.

Using this system, Fig. 1(a) is modified as shown in Fig. 1(b). Figure 1(b) indicates that the orthogonality between direct and delayed waves is kept and the direct and delayed waves of each channel are not contaminated by interference from the other channel within guard time.

3. Configuration of the proposed multi-code and multi-carrier hybrid transmision scheme

3.1 Configuration of the transmitter

Figure 4 shows a configuration of the transmitter. This proposed hybrid transmission scheme transmitter multiplexes the S/P converted data streams of each subchannel of OFDM by multi-code transmission using unique cyclic codes discussed in previous section. At the same time, in several selected subchannels of OFDM, the fixed pattern pilot data is inserted only in I-channel and spread by one of the cyclic modified M-sequences different from spread codes of traffic channels. This pilot signal is used to estimate frequency offset of OFDM transmission and delay profile characteristics in multipath fading environment. Then these spread sequence of all traffic channels and a pilot channel of all OFDM subchannels are modulated and multiplexed with OFDM.

3.2 Configuration of the receiver

At the receiver, as indicated in Fig.4, the received signal is sampled at the A/D converter and stored in the burst memory. The sampled data are fed into the frame timing estimator, which decides both the start points of modified M-sequence and DFT(Discrete Fourier Transform). We apply the feature of proposed cyclic modified M-sequence to estimate these points. As shown in Fig.5, the estimation method is to correlate between received data and delayed received data with the delay time of the number of the modified M-sequence without guard chips, accumulate these correlated results during successive symbols, integrate these accumulated data with the window of two times of guard chips and search the largest point of integrated value. The start points of DFT and cyclic code are decided by subtracting the number of the modified M-sequence without guard chips from this largest point. After decided the start point of DFT, the received data are fed into DFT circuit and detected as subchannel data.

The subchannel data are also detected by Rake reception of multi-code transmission. At first, each subchannel data are fed into the matched filter. In the matched filter, the subchannel data are correlated with the modified M-sequence without guard chips which spread the data of pilot channel. This complex delay profile, unfortunately, does not accurately estimate the propagation characteristics because of low its signal to noise power ratio (S/N). Therefore, we coherently sum the complex delay profiles measured in successive symbols. By using the data of measured propagation characteristics and the start point of modified M-sequence, we detect transmitted data of each OFDM channel with Rake receiver.

4. Computer simulation results

By using the code and chip timings recover scheme, proposed in above secion, The BER performance is discussed below with parameters indicated in Table 1 and cyclic modified M-sequence codes formatted in Fig. 3. These codes are designed to estimate waves delayed within 2 chips, and to RAKE diversity combine these estimated signals. We use 5-stage M-sequence (code length $2^5 - 1$), add the bias value and insert 2 guard chips before and after this sequence. Therefore the total code length is 35, and the maximum number of cyclic codes by one modified M-sequence is 10.

Figure 6 shows the BER performance when the number of multi-code transmission channels is 10 (include one pilot channel) per subchannel in static and one path Rayleigh fading environment under the both conditions that the code and chip timings are recovered perfectly and are estimated with the proposed joint estimation.

This figure indicates that the BER performance of proposed scheme is about 1dB worse than that of theoretical results. This reason is that in this proposed scheme we spread information signals by the code with the length of 35, however the process gain is 32 because our proposed codes add 4 guard chips and bias value to the normal 5 stage M-sequence. In addion, in 10 parallel transmission channels, one pilot channel is inserted. Furthermore, the BER performance in the double-spike Rayleigh fading environment with various Doppler frequencies is shown in Fig. 7 in the case of perfect synchronization. This figure indicates that the proposed scheme achieves good BER performance close to the theoretical value in fast Rayleigh fading channel.

5. Conclusions

We newly proposed a new Multi-carrier adn Multi-code hybrid transmission scheme that used cyclic code generated from a modified M-sequence to keep the orthogonality of codes not only between parallel channels but also between direct and delayed waves, and to reduce the number of codes used by a single user. The BER performance in Rayleigh fading envirnment was evaluated by computer simulations. Computer simulation results show that our proposed scheme achieves good BER performance in the fast selective Rayleigh fading environment and is suitable for future broadband land mobile communication systems.

References

[1] J.A.C. Bingham, "Multicarrier modulation for data transmission: An idea whose time has come," IEEE Commun. Mag., pp.5-14, May. 1990.

[2] C-L.I, G.P. Pollini, L. Ozarow and R.D. Gitlin, "Performance of Multi-Code CDMA Wireless Personal Communications Networks," in Proc. VTC'95, pp. 907-911, Jul. 1995.

[3] L. Vandendorpe, "Multitone Spread Spectrum Multiple Access Communications System in a Multipath Rician Fading Channel," in Proc. ICC'94, pp. 1638-1642, May. 1994.

[4] S.Kondo and L.B.Milstein, "Performance of multi-carrier DS CDMA Systems," IEEE Trans. Commun., vol.44, no.2, pp.238-246, Feb. 1996.

[5] H. Harada, G. Wu, K. Taira, Y. Hase and H. Sasaoka, "A New Multi-code High Speed Mobile Radio Transmission Scheme using Cyclic Modified M-sequence," in Proc. VTC'97, May. 1997.

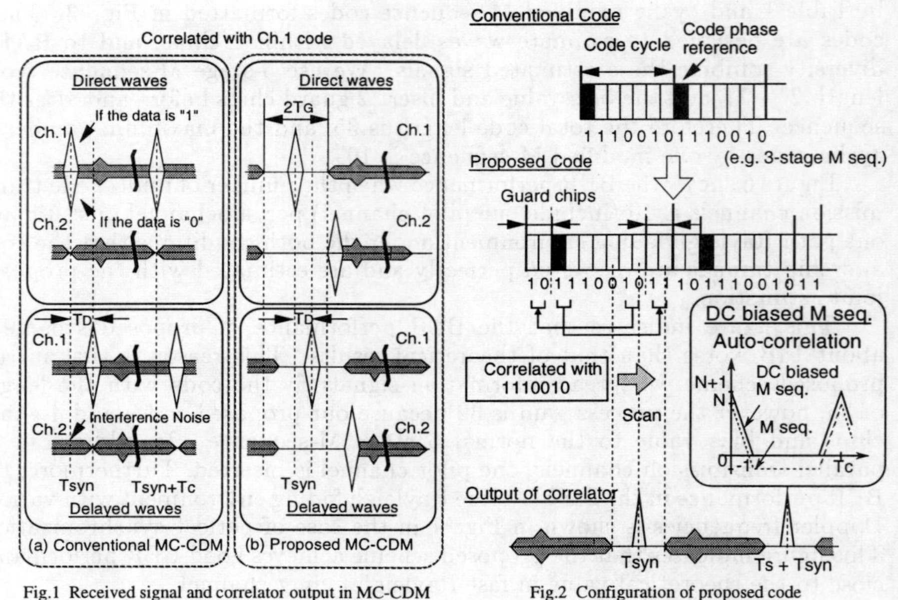

Fig.1 Received signal and correlator output in MC-CDM

Fig.2 Configuration of proposed code

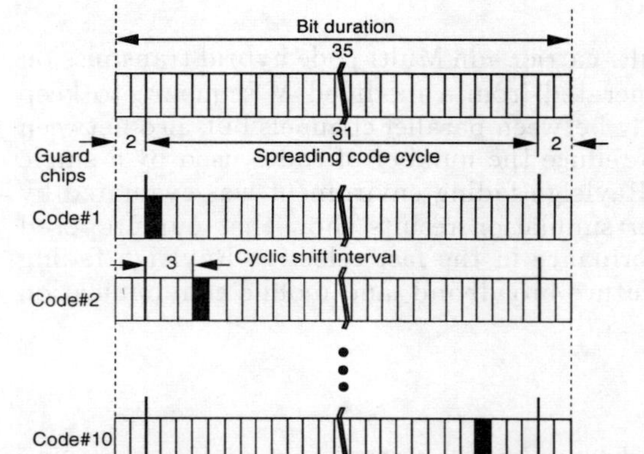

Fig.3 Example of proposed codes

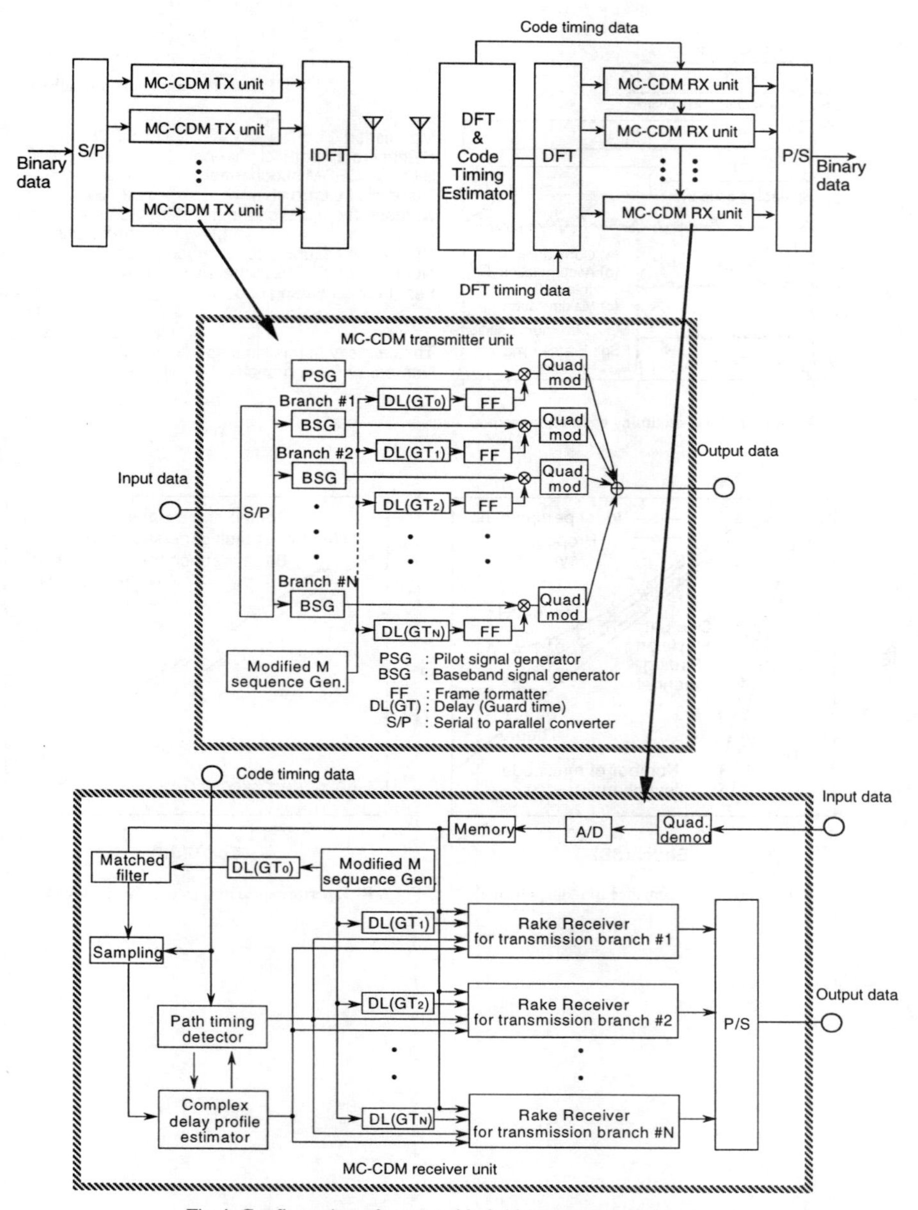

Fig.4 Configuration of proposed hybrid transmission scheme

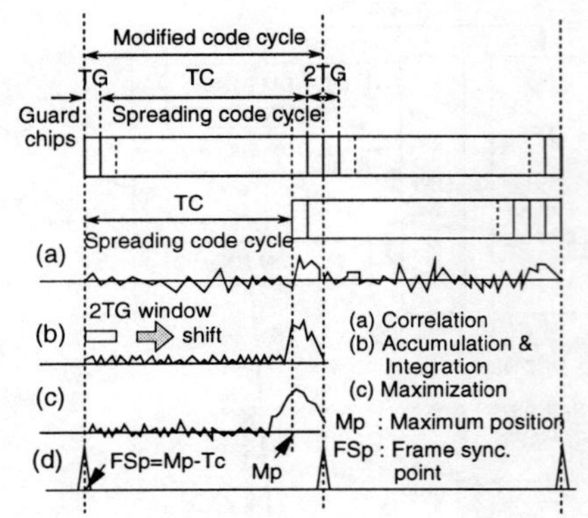

Fig.5 FFT and code timing estimation method

Table.1 Parameters used in this simulation

Modulation	: QPSK
Symbol rate / parallel channel	: 256ksymbol/s
Bit rate / OFDM subchannel	: 4608kbit/s
Number of Guard chip	: 2chip
Number of parallel channel	: 10ch (pilot 1ch traffic 9ch)
Number of accumulated symbols	: 20
Number of OFDM subchannels	: 16
Length of spreading code	: 31+2(before guard chips)+2(after guard chips)
The capacity of this simulation	: 73.728Mbps
Number of Rake branch	: 2

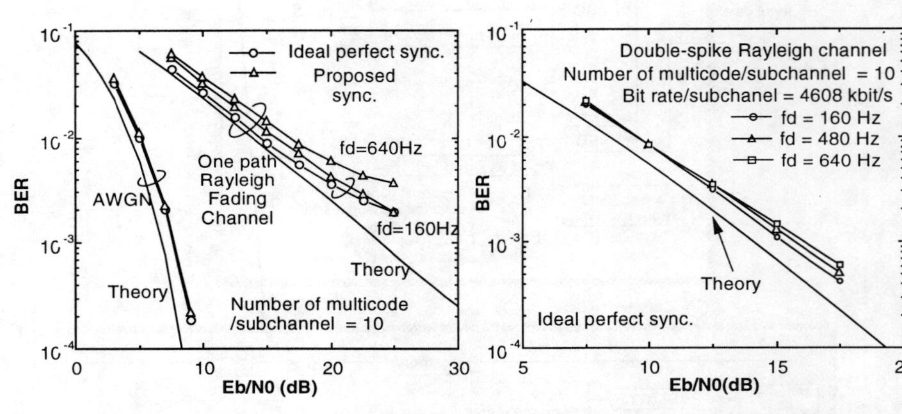

Fig.6 BER performance in fading channel

Fig.7 BER performance in selective fast fading channel

A SPREAD-SPECTRUM MULTI-CARRIER MULTIPLE-ACCESS SYSTEM FOR MOBILE COMMUNICATIONS

S. Kaiser and K. Fazel

Institute for Communications Technology
German Aerospace Research Establishment (DLR)
D-82234 Oberpfaffenhofen, Germany
E-mail: Stefan.Kaiser@dlr.de

ABSTRACT. A flexible multiple access scheme based on a novel combination of the spread-spectrum technique and orthogonal multi-carrier modulation is presented in this paper. The scheme is called spread-spectrum multi-carrier multiple access (SS-MC-MA). Its concept is similar to the concept of multi-carrier (MC)-CDMA. However, its basic difference is, that with SS-MC-MA the code division is used for simultaneous transmission of the data of a single user on the same sub-carriers, where with MC-CDMA, it is used for the simultaneous transmission of the data of different users on the same sub-carriers. The proposed scheme is investigated with channel coding and real channel estimation. SS-MC-MA achieves a high bandwidth efficiency for the up-link and for the down-link.

1 INTRODUCTION

Various performance comparisons between direct sequence (DS)-CDMA and multi-carrier (MC)-CDMA in the down-link of a mobile radio system demonstrate that the exploitation of orthogonal MC modulation can significantly

K. Fazel and G.P. Fettweis (eds.), Multi-Carrier Spread-Spectrum, 49-56.

increase the bandwidth efficiency of CDMA systems [1, 2, 3]. However, the application of MC-CDMA for the up-link has rarely been investigated, since the main problem encountered here, is the design of an efficient channel estimation. A novel multiple access scheme is proposed in this paper, which can achieve a bandwidth efficiency for the up- and the down-link comparable to the high bandwidth efficiency of MC-CDMA in the down-link, by guaranteeing a low complex channel estimation. The new scheme is called spread-spectrum multi-carrier multiple access (SS-MC-MA) and is also based on a combination of the spread-spectrum technique and MC modulation. However, its basic difference is that with SS-MC-MA the code division is used for simultaneous transmission of the data of a single user on the same sub-carriers, where in the case of MC-CDMA the code division is used for the transmission of the data of different users on the same sub-carriers. The concept of SS-MC-MA is similar to the concept of MC-CDMA with different groups of users [2, 4], however, with SS-MC-MA each group represents the data of another user and the multiple access is performed in the frequency domain. Therefore, no MAI is present within a cell. Due to the application of the spread spectrum technique, a data symbol is spread over the whole transmission bandwidth to exploit frequency diversity. Since all superimposed modulated spreading codes of one sub-set of sub-carriers are affected by the fading of the channel of only one user, even in the up-link, a simple channel estimation can be realized. For the joint detection of the data in one sub-set of sub-carries of a single user a low complex maximum-likelihood sequence estimation (MLSE) can be applied, similar to the low complex MLSE for the joint detection of the data of different users applied in an MC-CDMA system [4]. The system provides a high degree of flexibility due to variable sub-carrier allocation for variable data rates and can offer unequal error protection due to variable spreading code mapping.

2 SS-MC-MA CONCEPT

2.1 TRANSMISSION SCHEME

The block diagram of the SS-MC-MA transmitter (same for up- and down-link) for the data of user $i, i = 1 \dots N_u$, is shown in Fig. 1. N_u is the number of active users within a cell. After channel encoding, interleaving, and symbol mapping, L subsequent complex data-symbols $d_l^{(i)}, l = 1 \dots L$, of user i are serial to parallel converted. The vector $\mathbf{d}^{(i)}$ represents one block of L complex data-symbols of user i, $\mathbf{d}^{(i)} = (d_1^{(i)}, d_2^{(i)} \dots d_L^{(i)})^{\mathrm{T}}$, where $(.)^{\mathrm{T}}$ denotes the transposition. Each of the L data-symbols is multiplied with another orthogonal spreading code (e.g. Walsh-Hadamard code) of length L. The $L \times L$ matrix $\mathbf{C}$ represents the L orthogonal spreading codes. The modulated spreading codes are added symbol- and with that chip-synchronously, resulting in the transmission vector

$$\mathbf{s}^{(i)} = \mathbf{C} \cdot \mathbf{d}^{(i)} = (S_1^{(i)}, S_2^{(i)} \dots S_L^{(i)})^{\mathrm{T}}, \tag{1}$$

consisting of L chips. To increase the robustness of the SS-MC-MA scheme e.g. against inter cell interference, less than L modulated spreading codes of length L can be added to one transmission vector $\mathbf{s}^{(i)}$. The chips of M subsequent vectors $\mathbf{s}^{(i)}$ are frequency- and time-interleaved and fed to the OFDM unit, which also performs a user-specific frequency mapping.

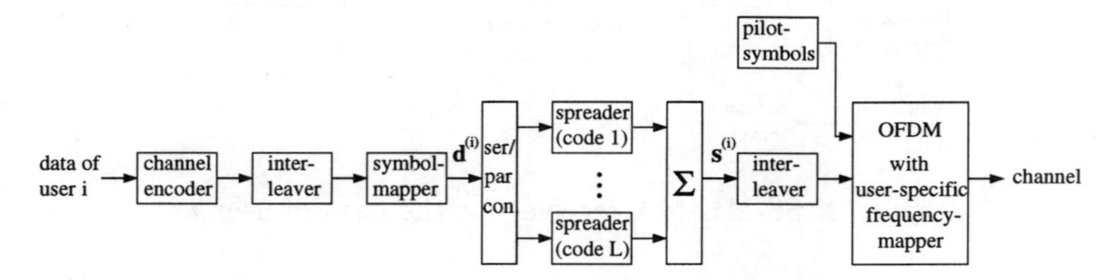

Figure 1: SS-MC-MA transmitter for the data of user i.

After the mapping, the chips $S_l^{(i)}, l = 1 \dots L$, are interpreted as complex frequency values of the sub-carriers. Each user of a cell transmits on a sub-set of sub-carriers, where the sub-sets of sub-carriers of the users within a cell are disjunct, according to an FDMA scheme. To exploit the diversity of the whole transmission channel, the specific sub-carriers assigned to a given user are distributed over the whole transmission bandwidth. Frequency diversity can be exploited if the distance between two sub-carriers assigned to a given user is higher than the channel coherence bandwidth. Finally, each OFDM-symbol of duration T_s is cyclic extended by a guard interval Δ to prevent inter symbol interference (ISI) in multipath channels [5]. In order to perform coherent detection at the receiver site and to guarantee robust synchronization (time and frequency), pilot-symbols are multiplexed in the data transmitted on each sub-carrier.

2.2 RECEPTION SCHEME

The block diagram of the SS-MC-MA receiver with coherent detection of the data of the i-th user is shown in Fig. 2. The received data are fed to the inverse OFDM unit which also performs the user specific frequency demapping and, furthermore, extracts the pilot-symbols from the data-symbols. The data-symbols are frequency and time deinterleaved, resulting in the received vector

$$\mathbf{r}^{(i)} = \mathbf{H}^{(i)} \cdot \mathbf{s}^{(i)} + \mathbf{n}^{(i)} = (R_1^{(i)}, R_2^{(i)} \dots R_L^{(i)})^{\mathrm{T}}. \tag{2}$$

The $L \times L$ diagonal matrix $\mathbf{H}^{(i)}$ describes the complex channel fading on the L sub-carriers of user i, and the vector $\mathbf{n}^{(i)}$ represents additive white Gaussian noise. The vector $\mathbf{r}^{(i)}$ is fed to the detection and symbol demapping unit. From the received pilot-symbols, the channel estimator estimates the channel state information (CSI), which is also frequency- and time-deinterleaved and used for the detection.

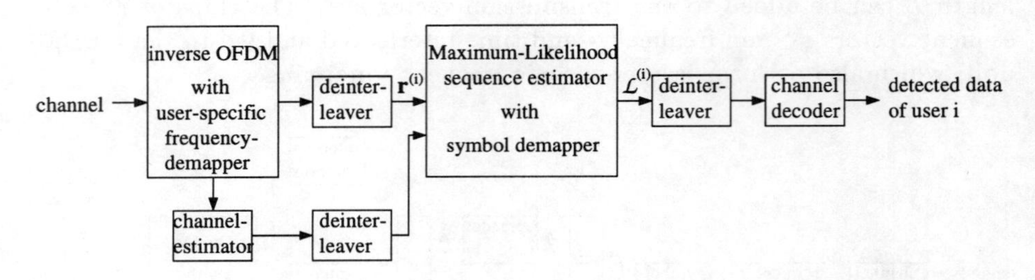

Figure 2: SS-MC-MA receiver for the data of user i.

An MLSE is applied for the detection of the data of a user. In the SS-MC-MA system we have the advantage that the MLSE in one estimation can decide on $2L$ code bits of a user if for instance QPSK mapping is used. Compared to MC-CDMA, no information about which users are active is required for MLSE with SS-MC-MA. Let us denote $^{+}\mathbf{S}$ a set of all possible transmitted sequences $\mathbf{s}_m$, where the considered code bit is equal to $+1$, and by $^{-}\mathbf{S}$ the set of all possible sequences, where the considered code bit is equal to -1, then, the log-likelihood ratio for MC-CDMA with MLSE results in [6]

$$\mathcal{L}^{(i)} = \ln \left(\frac{\sum\limits_{m=1, \mathbf{s}_m \in {}^{+}\mathbf{S}}^{2^L} \exp(-\frac{E_s}{N_0}\delta_m^2)}{\sum\limits_{m=1, \mathbf{s}_m \in {}^{-}\mathbf{S}}^{2^L} \exp(-\frac{E_s}{N_0}\delta_m^2)} \right) . \tag{3}$$

The Euclidean distance between the received sequence and the m-th possible transmitted sequence is denoted by δ_m^2. Finally, after de-interleaving, the channel decoder exploits the log-likelihood ratios, yielding the detected data bits of user i.

3 FLEXIBILITY OF SS-MC-MA

The proposed SS-MC-MA scheme inherently offers the following flexibilities:

- **Same radio interface for up- and down-link:** Since the SS-MC-MA scheme prevents MAI within a cell, we can use the same SS-MC-MA radio interface for the up-link as well as for the down-link.
- **Unequal error protection:** To increase the robustness of the SS-MC-MA scheme e.g. against inter cell interference, less than L modulated

spreading codes of length L can be added to one transmission vector $\mathbf{s}^{(i)}$, resulting in a higher redundancy due to spreading for a specific user.

- **Variable data rate due to variable sub-carrier allocation:** It is not necessary, that the transmission vector $\mathbf{s}^{(i)}$ of length L uses L sub-carriers per OFDM-symbol for its transmission. Since, in contrast to MC-CDMA, the chips of $\mathbf{s}^{(i)}$ contain the data of only one user, each user can transmit on a different number of sub-carriers.
- **Flexible TDMA component:** Since a synchronous transmission is assumed we can easily introduce a time division multiple access (TDMA) component in the OFDM-frame structure. With a TDMA component different groups of users can be supplied in subsequent OFDM-frames, resulting in an increasing number of active users.
- **Simple channel estimation for up- and down-link:** Since a sub-set of sub-carriers is exclusively used by one user, the channel fading on this sub-carriers can be estimated by a simple filtering in time direction on each sub-carrier. In the down-link the overhead due to pilot-symbols can drastically be reduced by filtering in time and frequency direction [7].

4 SYSTEM PERFORMANCE

The proposed SS-MC-MA scheme with L superimposed spreaded sequences implies that its performance is similar to the performance of MC-CDMA with full system load, when assuming perfect channel estimation. Thus, the assigned performance of SS-MC-MA for different detection techniques like minimum mean square error equalization, equal gain combining, or interference cancellation can be taken from [6, 8]. In the following, the up-link of a coded SS-MC-MA system with real channel estimation is considered. The results are also valid for the down-link, except that the redundancy due to the pilot-symbols required for channel estimation can be reduced from 20% to 5.6% in the down-link due to two-dimensional filtering [7].
The available transmission bandwidth is $B = 2$ MHz and the used carrier frequency is $f_c = 1.8$ GHz. By applying OFDM, the bandwidth B is split into $N_c = 256$ sub-carriers with a sub-carrier spacing of 7.8 kHz. The resulting OFDM-symbol duration is $T_s = 128\mu$s. The OFDM and inverse OFDM operation is realized by an IFFT and an FFT, respectively. A guard interval of $\Delta = 20\mu$s is inserted between adjacent OFDM-symbols to prevent ISI.
The SS-MC-MA system is based on a TDMA scheme, where, in a cycle of 4 frames (each frame consists of 31 OFDM-symbols, see Fig. 3), per frame another group of 32 users transmits. With this concept 128 active users can transmit with a net data rate of 10.46 kbit/s.

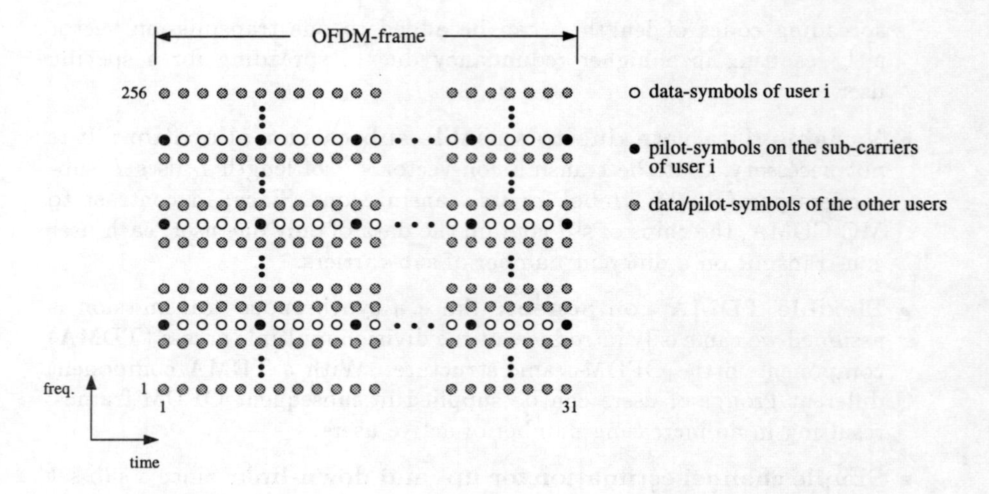

Figure 3: OFDM-frame with 256 sub-carriers and 31 OFDM-symbols.

The speech encoder of the mobile station delivers a data rate of 10.46 kbit/s to the channel encoder. The binary data are convolutionally encoded with rate 1/2 and memory 6. A block of 348 code bits is randomly interleaved and QPSK-mapped into 192 data-symbols. A block of 8 serial data-symbols is parallel converted and spread with orthogonal Walsh-Hadamard codes of length $L = 8$, where each of the 8 data-symbols is multiplied with another spreading code. The 8 data modulated spreading codes are added symbol-, and with that, chip-synchronously, resulting in a transmission sequence of 8 complex chips. A block of 192 chips of $M = 24$ subsequent transmission sequences are randomly frequency- and time-interleaved. In the OFDM unit 8 chips are modulated on 8 sub-carriers of an OFDM-symbol, where the user-specific sub-carriers are located in distances of 250 kHz, distributed over the whole bandwidth. On each of the 8 sub-carriers in a distance of 5 symbols a pilot-symbol is transmitted, which is known in the receiver (see Fig. 3).
The receiver applies an MLSE which detects separately in the I- and Q-components of the received data-symbols, using the estimated CSI. The CSI is estimated by Wiener-filtering (with 5 filter coefficients) of the received pilot-symbols on the user-specific sub-carriers over the time [9].
The performance of the SS-MC-MA system is investigated for two propagation scenarios defined in COST 207 [10], taking into account different velocities of the mobile users:

- **BU 3:** This channel model is valid for bad urban environments. The maximum delay is $10\mu s$. The velocity of the mobile user (pedestrian) is 3 km/h resulting in a maximum Doppler frequency of 5 Hz.

- **HT 100:** This channel model is valid for hilly terrain. The maximum delay is $20\mu s$. The velocity of the mobile user is 100 km/h resulting in a maximum Doppler frequency of 166.67 Hz.

In both channels, the Doppler spectrum is modeled as a Jakes spectrum [11]. The bit error rate (BER) in the up-link versus the average signal to noise ratio (SNR) E_b/N_0 of a SS-MC-MA system with real channel estimation (CE) is presented in Fig. 4. E_b is the average energy per received bit (data and pilot symbols are assumed to be transmitted with the same energy) and N_0 is the one-sided noise spectral density. As reference the performance of the SS-MC-MA system with perfect CE is given. The total SNR loss due to the guard interval Δ and the pilot symbols is 1.6 dB and is taken into account in the results of Fig. 4 (also in the curve with perfect channel estimation).

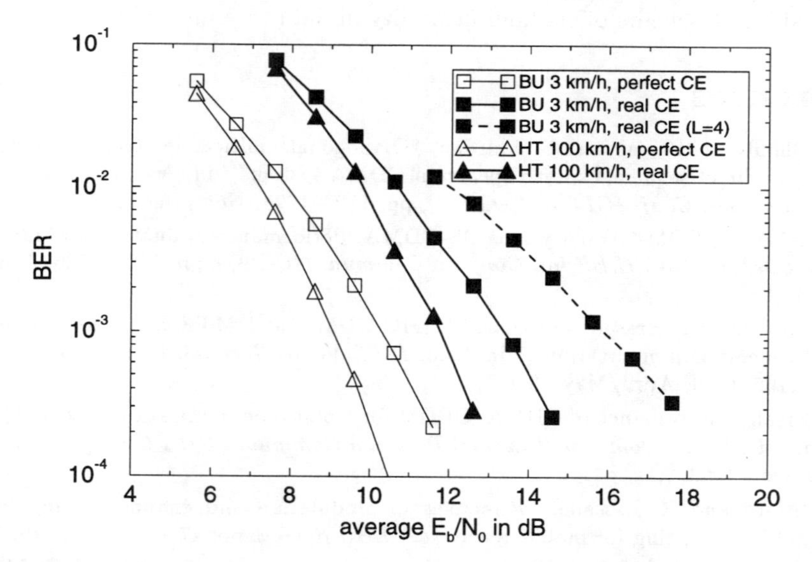

Figure 4: Performance of SS-MC-MA in different propagation scenarios: up-link, 128 active users, $B = 2$ MHz.

The performance is given for a full system load resulting in a bandwidth efficiency of 0.67 bit/s/Hz. Fig. 4 illustrates that the diversity offered with the HT 100 channel is 1.6 dB higher at a BER of 10^{-3} due to the increased delay spread and the increased Doppler spread compared to the BU 3 channel model. The degradation due to real channel estimation is about 2.5 dB - 3 dB at a BER of 10^{-3}. If the spreading code length is reduced from $L = 8$ to $L = 4$ (dashed curve), a performance degradation of 2.5 dB at a BER of 10^{-3} can be observed exemplary for the BU 3 channel.

5 CONCLUSIONS

A novel multiple access scheme has been presented based on the spread spectrum technique and orthogonal multi-carrier modulation (SS-MC-MA). The SS-MC-MA system is appropriate for the up-link as well as for the down-link of a mobile communications system by avoiding any MAI within a cell. The investigations take into account channel coding and real channel estimation by Wiener filtering. For the detection an MLSE with low complexity due to short spreading codes was used. The presented results show that SS-MC-MA in the up-link can, for instance, be applied in a hilly terrain (HT) where a mobile user has a velocity of 100 km/h @ 1.8 GHz with a SNR of 11.8 dB by guaranteeing a BER of 10^{-3} even with full system load. This yields a bandwidth efficiency of 0.67 bit/s/Hz, for the up-link as well as for the down-link. Furthermore, the new SS-MC-MA scheme offers high flexibility in data rate and robustness.

References

[1] A. Chouly, A. Brajal, and S. Jourdan, "Orthogonal multicarrier techniques applied to direct sequence spread spectrum CDMA systems," in *Proc. IEEE Global Telecommun. Conf. (GLOBECOM'93)*, pp. 1723–1728, Nov./Dec. 1993.

[2] S. Kaiser, "OFDM-CDMA versus DS-CDMA: Performance evaluation for fading channels," in *Proc. IEEE Int. Conf. on Commun. (ICC'95)*, pp. 1722–1726, June 1995.

[3] S. Hara and R. Prasad, "DS-CDMA, MC-CDMA and MT-CDMA for mobile multi-media communications," in *Proc. IEEE Vehic. Technol. Conf. (VTC'96)*, pp. 1106–1110, April/May 1996.

[4] K. Fazel, "Performance of CDMA/OFDM for mobile communication system," in *Proc. IEEE Int. Conf. on Universal Personal Commun. (ICUPC'93)*, pp. 975–979, Oct. 1993.

[5] M. Alard and R. Lassalle, "Principles of modulation and channel coding for digital broadcasting for mobile receivers," *EBU Review*, pp. 47–69, August 1987.

[6] S. Kaiser and L. Papke, "Optimal detection when combining OFDM-CDMA with convolutional and Turbo channel coding," in *Proc. IEEE Int. Conf. on Commun. (ICC'96)*, pp. 343–348, June 1996.

[7] P. Hoeher, S. Kaiser, and P. Robertson, "Two-dimensional pilot-symbol-aided channel estimation by Wiener filtering," in *IEEE Int. Conf. on Acoustics, Speech and Signal Processing (ICASSP'97)*, pp. 1845–1848, April 1997.

[8] S. Kaiser, "On the performance of different detection techniques for OFDM-CDMA in fading channels," in *Proc. IEEE Global Telecommun. Conf. (GLOBECOM'95)*, pp. 2059–2063, Nov. 1995.

[9] S. Haykin, *Adaptive Filter Theory.* Prentice Hall, 1986.

[10] Cost 207, "Digital land mobile radio communications," tech. rep., Luxembourg: Office for Official Publications of the European Communities, 1989.

[11] W. C. Jakes, *Microwave Mobile Communications.* John Wiley & Sons, 1974.

ORTHOGONAL FREQUENCY-DIVISION MULTIPLE ACCESS WITH FREQUENCY HOPPING AND DIVERSITY

Hikmet Sari

Alcatel Telspace
5, rue Noel-Pons
92734 Nanterre Cedex, France

ABSTRACT

We introduce the concept of orthogonal frequency-division multiple access (OFDMA) with frequency hopping and/or frequency diversity, and discuss its performance in the presence of narrowband interference which characterizes the return channel in cable TV (CATV) networks. In this scheme, rather than assigning a particular carrier frequency, the medium access control (MAC) protocol assigns a periodic sequence to the user which determines the carrier frequencies to be sequentially used by that user. We point out that with frequency hopping, OFDMA becomes a particular case of multicarrier code-division multiple access (CDMA), and we discuss the relationship of this technique with direct-sequence multicarrier CDMA which has become popular in the past few years.

INTRODUCTION

Orthogonal frequency-division multiple access (OFDMA) was proposed in [1] for the return channel in cable TV (CATV) networks. It derives from the multicarrier transmission technique known as orthogonal frequency-division multiplexing (OFDM) which has been adopted for digital terrestrial audio and video broadcasting [2], [3], and asymmetric digital subscriber lines (ADSL) [4]. In OFDM, different components of the multicarrier signal are generated locally, and are perfectly synchronized in terms of symbol timing and carrier frequency. In contrast, component signals in OFDMA are generated at different subscriber premises, and upstream modulators are simple single-carrier modulators which require some external sync information. Different user signals add up in the multipoint-to-point tree structure of the network, and the received signal at the cable head-end is an OFDM signal. In other words, OFDMA looks like a single-carrier system at the transmitter side and as an OFDM signal on the receiver side.

K. Fazel and G.P. Fettweis (eds.), Multi-Carrier Spread-Spectrum, 57-68.

OFDMA was one of the multiple access techniques investigated within the framework of the Digital Video Broadcasting (DVB) project conducted under the auspices of the European Broadcasting Union (EBU) for interactive services on coaxial or hybrid fiber/coax (HFC) cable networks. One of the major attributes of this technique is its robustness to narrowband interference. It is well known that code-division multiple access (CDMA) is superior to time-division multiple access (TDMA) in terms of robustness to narrowband interference, but as it turns out, OFDMA is still more robust than CDMA and is an attractive technique to carry upstream information from user premises to the head-end or to the fiber node in HFC networks. Following the adoption of TDMA by the Digital AudioVisual Council (DAVIC) [5], it was also decided to adopt this technique in the DVB Recommendations [6] for a European standard to be published by the European Telecommunications Standards Institute (ETSI). As it was pointed out in [7], TDMA is a widely used and well understood technique, and this fact obviously favors its adoption for new applications.

In this paper, we investigate new variants of OFDMA which employ frequency hopping, frequency diversity, or both. We discuss the relationship of this technique with multicarrier CDMA introduced in [8] and [9]. First, in the next section, we review the principle of OFDMA and its application to CATV networks. Next, we discuss the implementation issues of this technique including clock and carrier synchronization, power control, ranging, and the guard interval needed to cope with residual timing misalignments. Then, we generalize this technique to frequency hopping and frequency diversity, and discuss its relationship with multicarrier CDMA.

REVIEW OF OFDMA

To describe the principle of OFDMA, we consider the CATV network depicted in Fig. 1. This is a tree-and branch network emanating from a head-end and serving N subscribers. It is based on coax cable, but the discussion also holds for HFC networks. In this figure, we can distinguish a number of passive network elements (taps) which perform cable splitting functions as well as active network elements (amplifiers) which compensate for cable attenuation and power loss at bridge taps. Not shown in this figure are the upstream amplifiers needed to carry information from the subscribers to the head-end. Downstream information in CATV networks is typically transmitted in the frequency band of 120 to 860 MHz, and upstream information is transmitted in the frequency band of 5 to 50 Mhz.

Referring to Fig. 1, OFDMA consists of assigning carriers with a 1/T Hz frequency spacing to the different subscribers, where 1/T is the common symbol frequency. In other words, all clocks used at subscriber premises to generate the upstream data are locked to a common clock, and also the carrier spacing is controlled using the

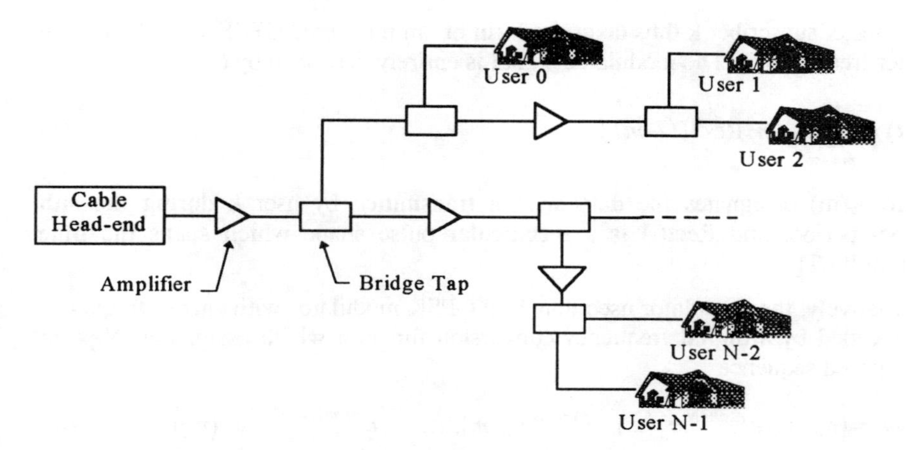

Fig. 1 : Typical CATV network

same clock signal. This synchronization is made possible by broadcasting pilot signals from the head-end, for instance, in the frequency band of 50 to 120 MHz which is used neither to broadcast data to user premises nor for upstream data transmission. From these pilot signals, upstream modulators derive their timing and carrier signals using a phase-lock loop (PLL). Since symbol timing and carrier frequencies are locally available at the head-end, they do not need to be recovered from the received OFDM signal. Carrier and timing recovery loops at the receiver are only needed for phase estimation. In other words, the OFDM receiver required at the cable head-end is a simple receiver which excludes some critical functions such as carrier frequency estimation.

In practical applications, the number of subscribers in the network is larger than the number of carriers available. Carriers are assigned to subscribers upon request, and this process is handled by the medium-access control (MAC) protocol. For simplicity, we will not discuss carrier assignment in this paper, and we will assume that the number of subscribers and the number of carriers are the same so that one carrier is assigned to each subscriber on a permanent basis, and consequently, we will not make any distinction between the words of subscriber and user. The signal transmitted by user k (k = 0,1,....., N-1) is of the form

$$s_k(t) = \mathrm{Re}\left\{a_k(t)e^{j2\pi f_k t}\right\} \tag{1}$$

with

$$f_k = f_0 + k/T. \tag{2}$$

Without any loss of generality, we assume in what follows that the modulation used for the return channel is quaternary phase-shift keying (QPSK). We also assume that the frequency assigned to the *k*th user is f_k. The upstream modulator at the premises of subscriber k thus takes the form of an unfiltered QPSK modulator with carrier frequency f_k. The modulated signal is entirely described by (1), (2), and

$$a_k(t) = \sum_{m=-\infty}^{+\infty} a_k(m) \mathrm{Rect}(t - mT) \tag{3}$$

where $a_k(m)$ designates the data symbol transmitted by user k during the *m*th symbol period, and Rect(t) is a rectangular pulse shape which spans the time interval [0, T[.

Alternatively, the modulator used may be a QPSK modulator with carrier frequency f_0, preceded by a digital frequency conversion function which assigns an N-point T/N-spaced sequence

$$A_k(m) = \left(a_k(m), e^{j2\pi k/N} a_k(m), e^{j4\pi k/N} a_k(m), \ldots\ldots, e^{j2\pi k(N-1)/N} a_k(m)\right) \tag{4}$$

to each $a_k(m)$ symbol of T-second duration. This sequence is easily generated using, for instance, a read-only memory (ROM) addressed with the symbol $a_k(m)$, the user index k, and a clock with frequency of N/T Hz. The reader will easily recognize that generation of the $A_k(m)$ sequence from the $a_k(m)$ symbol shifts the frequency spectrum by k/N Hz. This version of the modulator allows to use the same oscillator frequency f_0 for all modulators and generate individual carrier frequencies using a ROM.

At the cable head-end, the received signal is denoted r(t). It can be written as

$$r(t) = \sum_{k=0}^{N-1} s_k(t) + n(t) + I(t) \tag{5}$$

where n(t) denotes additive noise and I(t) is narrowband interference which characterizes typical CATV networks. This interference, which is also known as ingress noise, is a major obstacle to data transmission on the return channel of CATV networks [10]. After demodulating r(t) using a local oscillator with carrier frequency f_0, we obtain

$$x(t) = \sum_{k=0}^{N-1} x_k(t) + w(t) + J(t) \tag{6}$$

where $x_k(t)$ designates the complex envelope of the *k*th user signal $s_k(t)$, and w(t) and J(t) respectively designate baseband-equivalent noise and interference. Expression (6) can also be written as

$$x(t) = \sum_{k=0}^{N-1} e^{j2\pi(f_k - f_0)t} a_k(t) + w(t) + J(t) \tag{7}$$

where we can explicitly distinguish the information-bearing waveforms $a_k(t)$, k = 0,1,..., N-1.

The demodulated signal x(t) is sampled at the sampling rate of N/T Hz, and N regularly-spaced signal samples are thus generated per symbol period T. The N-point sequence generated by the sampler over the *m*th symbol period is given by

$$x_n(m) = \sum_{k=0}^{N-1} e^{j2\pi kn/N} a_k(m) + w_n(m) + J_n(m) \tag{8}$$

with n = 0,1,........, N-1. It is easily verified that except for a scaling factor of 1/N, the $x_n(m)$, n = 0,1,......., N-1, sequence is a noisy version of the inverse discrete Fourier transform (DFT) of $a_k(m)$, k = 0,1,......., N-1, which represents the sequence of symbols transmitted by the N users during the *m*th symbol interval. This indicates that the subscriber signals can be recovered using an N-point DFT after the sampler, and therefore, the receiver is an OFDM receiver. The block diagram of an OFDMA system is shown in Fig. 2.

IMPLEMENTATION ISSUES

In this section, we will discuss the basic issues related to the implementation of OFDMA and indicate how these problems can be handled. These include frequency synchronization of the oscillators used in the customer premises equipment (CPE) to generate upstream information, the ranging and power control processes, and the guard interval needed to cope with residual misalignment of different user signals.

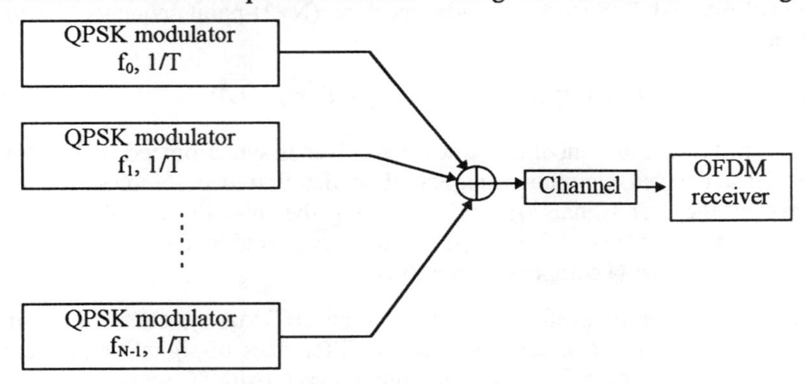

Fig. 1 : Typical CATV network

As mentioned earlier, OFDMA requires that the carrier spacing be identical to the symbol frequency that is common to all CPE modulators. In addition to constraining the upstream modulators to transmit at the same symbol frequency and fixing the carrier spacing equal to the symbol clock frequency, OFDMA also requires phase synchronism between clock frequencies of different CPE modulators. Without phase synchronism, orthogonality is lost between different carriers as it is illustrated in [1]. Since subscriber premises are at different distances from the cable head-end, ranging and power control are needed to phase-align and amplitude-balance different user signals. Note that power control and ranging are not specific to OFDMA, and are commonly used in conventional TDMA and CDMA systems. Power control and ranging are not only needed at installation, but they also need to be performed periodically in order to cope with eventual drifts which may be due to temperature variations and aging among other factors.

Unfortunately, phase-alignment of different user signals through ranging cannot be perfect, although it compensates for the major portion of delay differences. To compensate for residual misalignment, one can use the guard interval technique which consists of a cyclic extension of the symbol period. Without a guard interval, (4) describes the N-point T/N-spaced sequence generated by user k. The guard interval consists of extending the symbol period from T seconds to $T'=T+T_g$ seconds, where T_g is larger than the residual misalignment between different user signals. Suppose that the maximum delay difference between user signals is 2T/N seconds and that the extended symbol period is T′=T+2T/N. Referring back to (4), the (N+2)-point sequence that will be generated during the mth symbol period by the digital phase shifter preceding the modulator of user k will be

$$A'_k(m) = \left(a_k(m), e^{j2\pi k/N} a_k(m), \ldots, e^{j2\pi k(N-1)/N} a_k(m), a_k(m), e^{j2\pi k/N} a_k(m)\right). \quad (9)$$

Clearly, this extension gives two additional samples which are a repetition of the first and second samples of the original N-point sequence. Stated differently, the guard interval in OFDMA is a circular suffix (or prefix) to the N-point sequence per symbol interval T. Suppose that the received (N+2)-point sequence during the mth symbol interval is

$$X(m) = \left(x_0(m), x_1(m), x_2(m), \ldots\ldots, x_{N-1}(m), x_N(m), x_{N+1}(m)\right). \quad (10)$$

As shown in [1], if the symbol clock at the receiver is synchronized with the phase of the least delayed component signal, then the first two samples in (10) are corrupted by the user signals transmitted during the (m-1)th symbol, but all user signals, i.e., $a_0(m)$, $a_1(m)$,.........., $a_{N-1}(m)$ can be recovered by feeding to an N-point DFT operator the last N components of X(m).

More generally, a cyclic prefix of M points in an OFDMA system with N carriers and a symbol period T can cope with delay differences of signal components as long as the maximum difference does not exceed (M/N)T seconds. From the received N+M signal samples per symbol period, M points are dropped at the receiver, and the remaining N points are fed to the DFT operator to recover the N user symbols transmitted during that symbol period. The M symbols to be dropped are determined by the phase relation between the sampling clock at the receiver and the received individual user signals. The idea here resembles the guard interval issue in OFDM systems, but the guard interval does not play the same role in both systems : Whereas the cyclic prefix compensates for delay differences of individual signal components in OFDMA, it is needed to cope with multipath propagation and transform the linear convolution of the channel into a circular convolution in OFDM [11].

COMPARISON WITH TDMA AND CDMA

At this point, it is instructive to discuss the relative merits of the three basic multiple access techniques which are TDMA, CDMA, and (O)FDMA. Both TDMA and classical FDMA have been in use for decades and are now well understood by the technical community. Furthermore, FDMA is generally perceived as an old-fashioned technique, and the competition today is often between TDMA and CDMA in many communication system designs. Despite this, FDMA is still a basic ingredient of any system, because neither in TDMA nor in CDMA, it is desirable to transmit very wideband single-carrier signals. As a result, TDMA systems in use today are in fact hybrid TDMA/FDMA, and CDMA systems are in fact hybrid CDMA/FDMA.

Further, whether the system is designed for fixed or mobile applications, and whether it is satellite-based or terrestrial, one can read very controversial statements concerning the comparison of TDMA and CDMA in terms of system capacity, but these statements are technically questionable, and it is not unfair to say that they are often driven by commercial motivations rather than based on solid technical arguments. We will not enter this discussion here, but we will simply indicate that spread-spectrum signaling which forms the basis of CDMA is efficient against narrowband interference, and CDMA is therefore more robust than TDMA in the presence of this type of interference. If the spectral spreading factor is γ, the gain of CDMA with respect to TDMA in terms of narrowband interference is $\Gamma=10.\text{Log}_{10}(\gamma)$. There is a threshold effect, however, both in TDMA and CDMA : If we investigate the number of users who can communicate reliably (say at a bit error rate lower than 10^{-3}) over the channel, this number suddenly goes from N to 0 as the interference level exceeds some threshold.

OFDMA enters this race as the newest multiple access technique, although many of its features are those of conventional FDMA. What truly differs OFDMA from conventional FDMA is that adjacent signal spectra may overlap and that a single demodulator is sufficient to accommodate N simultaneous users. In fact, FDMA is by far more efficient that both TDMA and CDMA in terms of robustness to narrowband interference. To be specific, suppose that the interference is limited to p adjacent carrier frequencies. An FDMA system will still be able to accommodate N-p users independent of the interference level, whereas TDMA and CDMA systems will break as the interference level exceeds some threshold. OFDMA has a similar behavior, but the number of unusable frequencies is not strictly limited to p, because of the overlap between adjacent carrier spectra. Despite this, OFDMA is substantially superior to TDMA in terms of robustness to continuous-wave (CW) interference, as it was demonstrated in [1]. Also, some more recent work by other authors [12] showed that, as expected, OFDMA is also substantially superior to CDMA in this environment.

OFDMA WITH FREQUENCY HOPPING AND DIVERSITY

In the basic version of OFDMA, each carrier is assigned to one user. It is not difficult to figure out that this multiple access technique easily lends itself to frequency hopping. Rather than assigning a particular frequency to each user, the cable head-end can assign a hopping sequence. One of the simplest frequency hopping techniques consists of cyclic linear shifting of the frequency assigned to each user over the total signal bandwidth. To visualize the operation of this technique, we will consider a 512-carrier OFDMA system. During the first symbol period, the first carrier is assigned to the first user, the second carrier to the second user, and so forth. With a 1-step frequency shift between the first and the second symbols, the first user will have the second carrier, the second user will have the third carrier, and so forth, during the second symbol period. Similarly, an additional 1-step frequency shift is applied to all users between the second and the third symbol period. This frequency assignment procedure can be mathematically formulated as follows.

$$f(k,m) = f_{k+m(\mathrm{mod}.512)} \tag{11}$$

where f(k,m) designates the frequency assigned to the *k*th user during the *m*th symbol period, and both indexes k and m are defined over the interval 0,1,.....,511. One carrier frequency is thus permanently assigned to each user, but the carrier frequency cyclically sweeps the signal bandwidth.

In this frequency hopping scheme, all users make use of all carrier frequencies, but during different symbol periods. In the presence of a stationary narrowband interference, all users will have the same performance, and the symbols transmitted over carriers corrupted by interference will have a high error probability. Coding becomes necessary to protect these symbols exactly in the same way that coding is necessary in OFDM systems to operate on frequency-selective fading channels. If the error correction code used is a block code, one interesting choice is to select the block size in such a way that it coincides with one cycle of the periodic hopping sequence.

In the example chosen, the number of bits transmitted by each user over a period of the periodic frequency hopping sequence is 1024 (with QPSK modulation). This corresponds to 128 8-bit bytes. Suppose now that the error correction code used is a shortened Reed-Solomon (RS) code [13] defined over the Galois field GF(2^8), i.e., the code symbols are 8-bit bytes, and suppose further that it is specified to correct 4 bytes per block of 128 bytes. The number of redundant bytes per block in this coding scheme is 8, which corresponds to a bandwidth expansion of 6.2 %. It is clear that the described coded frequency hopping scheme can cope with narrowband interference which affects up to 16 carrier frequencies. In fact, if the OFDMA system uses strictly bandlimited and nonoverlapping frequency pulses, it would entirely reject a narrowband interference spanning 16 carrier frequencies, but OFDMA systems with unfiltered QPSK (or any other linear modulation) lead to

Sinc(x)=Sin(x)/x shaped frequency pulses, and the transmitted signal spectrum centered on the carrier frequency f_k is of the form

$$S_k(f) = Sinc^2\left[\pi T(f - f_k)\right]. \tag{12}$$

Since adjacent frequency pulses overlap, a narrowband interference with frequency bandwidth limited to n/T corrupts more than n adjacent frequency pulses, but the described scheme can still handle an interferer spanning several carrier frequencies.

Next, consider the same frequency-hopped OFDMA scheme, but suppose that it uses a convolutional code rather than a block code. In that case, error events will frequently happen when the successive carriers assigned to the user fall within or in the vicinity of the interferer frequency. This phenomenon causes frequent error events and seriously affects the quality of service. Fortunately, the frequency hopping procedure can be easily changed to retain the system resistance to narrowband interference. The key is to increase the frequency step from one symbol to another so that interference will typically not affect several successive symbols transmitted by one user. This is equivalent to combining the 1-step frequency hopping scheme with a periodic block interleaver. If the frequency step is l such that ln=512, one period of the periodic frequency hopping scheme for user k is described by the frequency index array

$$\begin{array}{lllll}
k, & k+l, & k+2l, & k+3l, \ldots\ldots, & k+(n-1)l \\
k+1, & k+l+1, & k+2l+1, & k+3l+1, \ldots\ldots, & k+(n-1)l+1 \\
k+2, & k+l+2, & k+2l+2, & k+3l+2, \ldots\ldots, & k+(n-1)l+2 \\
\ldots\ldots & & & & \\
k+l-1, & k+2l-1, & k+3l-1, & k+4l-1, \ldots\ldots, & k+nl-1.
\end{array} \tag{13}$$

If l is larger than the length of most likely error events of the convolutional code used, the frequency hopping scheme will be efficient against the interferer. As an example, suppose that the most likely error events of the convolutional code have a length less than 8 and that the frequency hopping parameter l is chosen as 8 (or higher). In this example, only one sample corrupted by the narrowband interferer will be contributing to the current receiver decisions.

OFDMA WITH FREQUENCY DIVERSITY

The frequency hopping and error correction coding schemes discussed in the previous subsection introduce some form of frequency diversity in the transmitted signal. Indeed, since detection is made block by block or by sequence, the receiver decision on symbols transmitted at carrier frequencies affected by narrowband interference will benefit from the higher reliability of symbols transmitted at other carrier frequencies. OFDMA also easily lends itself to explicit frequency diversity which consists of transmitting the same information at two or more different frequencies. If the frequency spacing between carriers assigned to one user is much larger than the interferer bandwidth, then only one of these signals will be affected, and the information symbol can be reliably extracted from the other (redundant) carrier(s). A simple example of this technique is an OFDMA scheme in which N/2 users share N carriers according to the rule that both f_k and $f_{k+N/2}$ are assigned to user k, where the k index spans the interval (1, 2, ..., N/2). It is clear that in the presence of narrowband interference, one of the two samples assigned to each user will be unaffected by the interferer, and all user signals will be reliably detected.

This is a true diversity in the traditional sense which doubles the spectral occupancy of the transmitted signal, but in principle, it requires no error correction coding and gives a higher robustness against narrowband interference as compared to implicit frequency diversity which consists of frequency hopping combined with low-redundancy error-correction coding. Frequency diversity can also be combined with frequency hopping and error correction coding, but we will not further discuss this technique which makes an inefficient use of the available spectrum.

DIRECT-SEQUENCE CDMA

At this point, it becomes clear that with frequency hopping, OFDMA coincides with multicarrier FH-CDMA, and it is instructive to clarify its relationship with multicarrier DS-CDMA. First, note that direct-sequence spectral spreading (DS-SS) is a kind of repetition coding, and in multicarrier DS-SS, the information symbols are repeated at different carrier frequencies rather than in time. Therefore, multicarrier DS-CDMA can be viewed as an OFDMA system with N-fold frequency diversity, where N is the number of carriers. If we focus on Walsh-Hadamard sequences [14], there exists N orthogonal sequences which can be assigned to N different users to share the same channel bandwidth. In other words, this CDMA technique can accommodate the same number of users as TDMA and FDMA on an additive white Gaussian noise (AWGN) channel. Let P_k , k = 0, 1, ..., N-1 designate the N orthogonal Walsh-Hadamard sequences of length N. They can be written as

$$P_k = \left(p_{k,0}, p_{k,1}, \ldots\ldots\ldots\ldots, p_{k,N-1}\right). \qquad (14)$$

Recall that in the basic OFDMA scheme without spectral spreading, (4) gives the N-point sequence generated by the modulator of user k during the *m*th symbol period. That is, the user and the carrier indexes coincide. In OFDMA with Walsh-Hadamard spectral spreading, the sequence generated by the upstream modulator of user k is

$$B_k(m) = \sum_{n=0}^{N-1} p_{k,n}\left(a_k(m), e^{j2\pi n/N} a_k(m), e^{j4\pi n/N} a_k(m), \ldots\ldots\ldots, e^{j2\pi(N-1)/N} a_k(m)\right)$$

$$= a_k(m)\sum_{n=0}^{N-1} p_{k,n}s_n \tag{15}$$

with

$$s_n = \left(1, e^{j2\pi n/N}, e^{j4\pi n/N}, \dots\dots\dots\dots\dots\dots, e^{j2\pi(N-1)/N}\right). \tag{16}$$

In this scheme, each user permanently occupies the total signal bandwidth. The information sequence of a user is repeated N times over the channel bandwidth and multiplied by the Walsh-Hadamard sequence assigned to it. The original sequence is recovered at the receiver by multiplying the received sequence by the same Walsh-Hadamard sequence.

DISCUSSION AND CONCLUSIONS

We have presented some new variants of the OFDMA technique presented earlier in [1]. This includes OFDMA with frequency hopping, frequency diversity, or both. With frequency hopping, OFDMA coincides with multicarrier CDMA in which spectral spreading is based on frequency hopping. It is also interesting to note that the OFDMA approach leads to multicarrier DS-CDMA when it employs full frequency diversity and Walsh-Hadamard sequences to separate different user signals. These observations are not surprising, because multicarrier DS-CDMA is based on spreading the spectrum of the transmitted signal over the N carriers in the channel bandwidth, and this process is performed in the frequency domain, as opposed to spreading in the time domain in conventional CDMA systems.

We have not investigated the relative performance of OFDMA with frequency hopping with respect to the basic OFDMA scheme and to TDMA, and this can be pointed out as a topic for further study. It can be conjectured, however, that OFDMA with frequency hopping behaves similarly to CDMA, i.e., it outperforms TDMA, but does not achieve the performance of the original OFDMA. This implies that in (O)FDMA, it is better to avoid using carrier frequencies corrupted by narrowband interference rather than distributing the data stream over the entire channel bandwidth and using error correction coding and interleaving to improve the reliability of the symbols transmitted on those frequencies. This conjecture is supported by the following heuristic arguments assuming upstream data transmission employs a block code and hard-decision decoding. In this case, the block code involves 2p redundancy symbols per block in order to correct p symbol errors, which indicates that in order to cope with a narrowband interferer which affects p adjacent carrier frequencies, 2p symbols per block must be sacrificed. That is, the redundancy of the code largely outweighs the gain associated to the use of frequencies affected by the interference.

These heuristic arguments tend to point out that in the presence of narrowband interference, the basic OFDMA scheme without frequency hopping is the most robust multiple access technique, provided that carrier frequencies affected by interference can be identified and discarded from the assignment procedure. This is not difficult to implement if the interference is time-invariant, but in the presence of time-varying interference, it requires periodically measuring the channel frequency response, updating the carrier frequency assignment in accordance with the new response of the channel, and possibly retransmitting some data blocks corrupted by interference.

REFERENCES

[1] H. Sari, Y. Levy, and G. Karam, "Orthogonal Frequency-Division Multiple Access for the Return Channel on CATV Networks," ICT '96 Conf. Rec., vol. 1, pp. 52-57, April 1996, Istanbul.

[2] M. Alard and R. Lassalle, "Principles of Modulation and Channel Coding for Digital Broadcasting for Mobile Receivers," EBU Review, no. 224, pp. 3-25, August 1987.

[3] B. Marti et al., "European Activities on Digital Television Broadcasting : From Company to Cooperative Projects," EBU Technical Review, no. 256, pp. 20-29, 1993.

[4] P. S. Chow et al., "A Multicarrier E1-HDSL Transceiver System with Coded Modulation," European Trans. Telecommunications, vol. 4, no. 3, pp. 257-266, May-June 1993.

[5] DAVIC 1.0 Specifications : Cable Modem Baseline Document, Revision 2.0, June 1996.

[6] prETS 300 800 : "Digital Video Broadcasting (DVB) : DVB Interaction Channel for Cable TV Distribution Systems," TM 1640 rev. 4, June 1996.

[7] D. D. Falconer, F. Adachi, and B. Gudmundson, "Time Division Multiple Access Methods for Wireless Personal Communications," IEEE Communications Magazine, vol. 33, no. 1, pp. 50-57, January 1995.

[8] N. Yee, J.-P. Linnartz, and G. Fettweis, "Multicarrier CDMA for Indoor Wireless Radio Networks," Proc. PIMRC '93, pp. 103-113, Sept. 1993, Yokohama, Japan.

[9] K. Fazel and L. Papke, "On the Performance of Convolutionally Coded CDMA/OFDM Mobile Communications Systems," Proc. PIMRC '93, pp. 468-472, Sept. 1993, Yokohama, Japan.

[10] C. A. Eldering, N. Himayat, and F. M. Gardner, "CATV Return Path Characterization for Reliable Communications," IEEE Communications Magazine, vol. 33, pp. 62-69, August 1995.

[11] H. Sari, G. Karam, and I. Jeanclaude, "Transmission Techniques for Digital Terrestrial TV Broadcasting," IEEE Communications Magazine, vol. 33, pp. 100-109, February 1995.

[12] M. van Bladel and M. Moeneclaey, "Comparison of Access Techniques for the CATV Return Channel," Proc. IEEE 4th Symposium on Communications & Vehicular Technology in the Benelux, pp. 27-35, October 1996.

[13] G. C. Clark, Jr., and J. B. Cain, Error-Correction Coding for Digital Communications, New York : Plenum Press, June 1981.

[14] N. Ahmed K. R. Rao, "Orthogonal Transforms for Digital Signal Processing," Berlin Heidelberg : Springer Verlag, 1975.

M-ARY ORTHOGONAL MODULATION FOR MC-CDMA SYSTEMS IN INDOOR WIRELESS RADIO NETWORKS

Armin Dekorsy and Karl-Dirk Kammeyer
University of Bremen, FB-1, Department of Telecommunications
P.O. Box 33 04 40, D-28334 Bremen, Germany,
Fax: +(49)-421/218-3341, e-mail: dekorsy@comm.uni-bremen.de

ABSTRACT

In this paper, we introduce the orthogonal M-ary Walsh modulation for *multicarrier code-division-multiple-access* (MC-CDMA) transmission. Analytical statements will be given for an AWGN channel and simulation results will be illustrated for a Rayleigh fading indoor channel assuming uplink transmission and coherent reception. Comparisons with antipodal modulation show that Walsh modulation performs better in terms of bit error rate and spectral efficiency.

1. INTRODUCTION

In mobile radio communication systems, much attention has been paid to CDMA. One well-known scheme is the so-called QUALCOMM system (IS-95) combining *direct sequence* CDMA (DS-CDMA) with M-ary orthogonal Walsh modulation. Results illustrated in [1] show the performance improvement of

K. Fazel and G.P. Fettweis (eds.), Multi-Carrier Spread-Spectrum, 69-76.

Walsh modulation over classical modulation schemes such as BPSK.
Recently, MC-CDMA has been proposed combined with classical modulation [2, 3, 4, 5] and performances of different detection strategies have been analysed. It has been shown that MC-CDMA has a better spectral efficiency compared to DS-CDMA [6].
The aim of this paper is to introduce the M-ary Walsh modulation technique for MC-CDMA transmission. In particular, the performance of an AWGN channel will be evaluated analytically. Simulation results will be shown for the uplink bit error rate (BER) for coherent reception assuming a Rayleigh fading indoor radio channel. The results are also contrasted with BPSK performance.

2. TRANSMISSION SYSTEM

With CDMA systems in general, the *multiple access interference* (MAI) is mainly determined by the codes implemented. For MC-CDMA transmission over a downlink channel, where we have a synchronous situation, optimized codes like Walsh codes can be used. Since every user-specific codeword is affected by the same channel transfer function, the deterioration of the correlation properties can be restored by using suitable equalization schemes. For an uplink transmission scenario, which is considered in this paper, every user-specific codeword is affected by different channel transfer functions. This fact leads to an asynchronous situation involving the use of PN-sequences, which results in higher MAI . Furthermore, *maximum ratio combining* (MRC) detection strategy is applied, being proper to reduce the interferences caused by the frequency selective fading of the channel [7]. Ideal channel estimation is assumed and no channel coding is involved.
In contrast with single-carrier transmission, a MC-CDMA scheme yields a frequency resolution by dividing the available bandwidth into a finite number of subbands. The idea behind the application of M-ary Walsh modulation is to further increase this resolution in order to distinguish transmitted symbols without raising the required bandwidth. Therefore, instead of spreading one data symbol over a number of subcarriers, spreading can be achieved in two successive steps. By applying Walsh modulation, we further use orthogonality, which renders good performance in general. Such a MC-CDMA transmitter (mobile station) involving Walsh modulation is illustrated in fig. 1. For simplicity, one of J active users is taken into account (subscripts are omitted). The data bits, $b \in \{0, 1\}$ each of duration T_b, are serial/parallel converted to groups of $\log_2(M)$ data bits each. The Walsh modulation maps the $\log_2(M)$ data bits to one corresponding Walsh symbol (vector) $\boldsymbol{w}^m = [w_0^m, w_1^m, \ldots, w_{M-1}^m]^T$, $m \in \{0, M-1\}$ including M Walsh chips $w_\mu^m \in \{-1, 1\}$, $\mu = 0 \ldots M-1$. Each of the M parallel Walsh chips has a duration of $T = \log_2(M)\, T_b$. This modulation can also be interpreted as frequency spreading of value $M/\log_2(M)$. The set of M orthogonal Walsh symbols represents an orthogonal basis which can

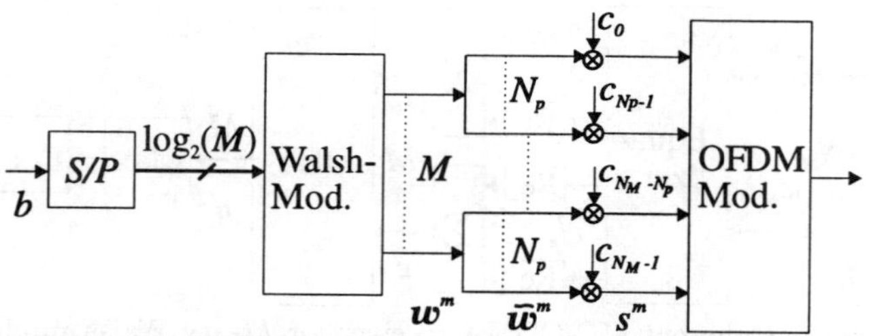

Figure 1: MC-CDMA transmitter with M-ary Walsh modulation

be evaluated recursively by applying

$$\boldsymbol{W}_{2M} = \begin{pmatrix} \boldsymbol{W}_M & \boldsymbol{W}_M \\ \boldsymbol{W}_M & -\boldsymbol{W}_M \end{pmatrix}, \quad \boldsymbol{W}_1 = (1), \tag{1}$$

where $\boldsymbol{W}_M$ is the $M \times M$ *Hadamard-Matrix*. The Euclidian distance is identical for all possible pairs of symbols and equals

$$\varepsilon = \sum_{\mu=0}^{M-1} |w_\mu^m - w_\mu^n|^2 = 2M \quad \forall \quad m \neq n. \tag{2}$$

To finally get the transmitted symbols, the Walsh chips are replicated into N_p parallel copies where each branch of the parallel stream is multiplied with one chip of the user specific code $c_i \in \{\pm 1/\sqrt{N_M}\}$, $i = 0 \ldots N_M - 1$. Since each bandwidth of the M parallel subchannels decreases for larger values of M, the number of subcarriers N_M increases for the same available bandwidth B. The number of subcarriers is determined by $N_M = N_2 \cdot \log_2(M) = M \cdot N_p$, where N_2 is the number of subcarriers used for 2-ary Walsh modulation or BPSK, respectively. For the jth user, this yields the transmission vector $\boldsymbol{s}^{m,j}$ before the OFDM modulation

$$\boldsymbol{s}^{m,j} = [s_0^{m,j} \ldots s_{N-1}^{m,j}]^T, \quad s_i^{m,j} = \tilde{w}_i^{m,j} \cdot c_i^j \tag{3}$$

with $\tilde{w}_i^{m,j} = w_\mu^{m,j}$, $\forall\ \mu = \lfloor i/N_p \rfloor$, m, j. OFDM modulation includes the IFFT (IDFT) and inserts the guard interval between adjacent OFDM symbols. To prevent *intersymbol interference* (ISI) and *adjacent channel interference* (ACI), the guard time T_g is chosen such that $T_g \geq \tau_{max}$, where τ_{max} is the maximum delay spread of the channel. Since the channel is regarded slowly time selective, $T \ll 1/f_{d,max}$ holds and $f_{d,max}$ indicates the maximum Doppler frequency. Due to the insertion of the guard interval, every subcarrier is affected by the corresponding channel transfer coefficient. For the uplink transmission scenario assumed here, the ith coefficient for every user j is given by $h_i^j = \rho_i^j \cdot \exp(\phi_i^j)$, where ρ_i^j and ϕ_i^j are the random amplitude and phase. Hence, the channel can be described for every user by an $N_M \times N_M$ diagonal matrix $\boldsymbol{H}^j$.

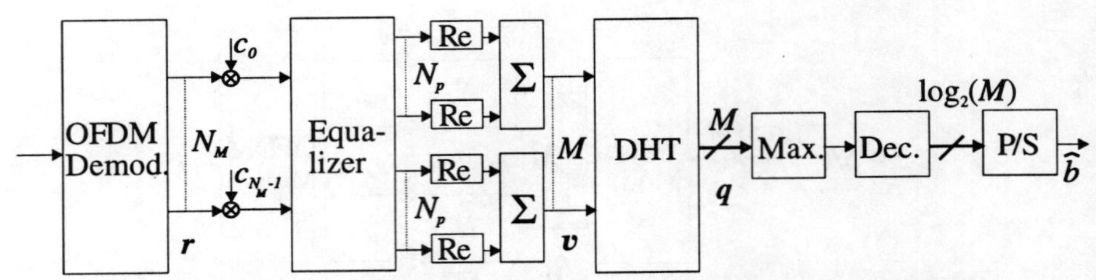

Figure 2: Coherent MC-CDMA receiver for M-ary Walsh modulation

The coherent MC-CDMA receiver for M-ary Walsh modulation is presented in fig. 2. Paying attention to J active users, the received signal after OFDM demodulation can be written as a sum of vectors $\boldsymbol{r}^{m,j}$

$$\boldsymbol{r} = \sum_{j=0}^{J-1} \boldsymbol{r}^{m,j} + \boldsymbol{n} = \sum_{j=0}^{J-1} \boldsymbol{H}^j \boldsymbol{s}^{m,j} + \boldsymbol{n} \tag{4}$$

with elements $r_i^{m,j} = h_i^j \cdot s_i^{m,j}$, $i = 0 \ldots N_M - 1, m \in \{0, M-1\}$. The vector $\boldsymbol{n}$ represents AWGN. After multiplication with the user specific code and equalization, the first part of despreading is obtained by subcorrelating N_p subcarriers. Reception for the user $j = 0$ is assumed (subscripts are omitted). The components of $\boldsymbol{v}$ are given by

$$v_\mu = \sum_{i=\mu N_p}^{(\mu+1)N_p-1} \mathrm{Re}\left\{ c_i e_i r_i^{m,0} + \sum_{j=1}^{J-1} c_i e_i r_i^{m,j} + c_i e_i n_i \right\}, \quad \mu, m \in \{0, M-1\} \tag{5}$$

where e_i indicates the equalization coefficient of the ith subcarrier. Since MRC is considered $e_i = h_i^*$. To obtain *maximum likelihood detection* (MLD), the signal is correlated with all possible Walsh symbols $\boldsymbol{w}^m, m \in \{0, M-1\}$. The MLD can be realized by the *Discrete Hadamard Transform* (DHT) [1], using the Hadamard matrix eq.(1) according to

$$\boldsymbol{q} = \boldsymbol{W}_M \cdot \boldsymbol{v}. \tag{6}$$

Vector $\boldsymbol{q}$ includes M decision variables where the choice of the maximum value results in a symbol decision. Decoding of the decided symbol to $\log_2(M)$ data bits and parallel/serial conversion leads finally to the estimated data stream with bits $\hat{b}$.

3. AWGN CHANNEL

For a large number of users, the users can be considered as real Gaussian noise. In the case of random codes, the interference power equals $\beta_{int}\, E_b \approx (J-1)/N_M\, E_b$, where E_b is the data bit energy[1]. Assuming coherent reception,

[1] $E_b = 0.5A^2\, T_b = 0.5 T_b$ where A indicates the amplitude of the symbol.

the bit error probability P_b of M-ary Walsh modulation is given by [1]

$$P_b = A_M \int_{-\infty}^{\infty} [1 - [1 - \frac{1}{2}\mathrm{erfc}(x)]^{M-1}] \exp\left[-\left(x - \sqrt{\log_2(M)\,\gamma}\right)^2\right] dx \tag{7}$$

with

$$A_M = \frac{M/2}{\sqrt{\pi}(M-1)}, \quad \gamma = S/N \cdot \frac{1}{1 + 2\beta_{int} S/N \log_2(M)}. \tag{8}$$

S/N is the signal-to-noise ratio used for MLD. The guard interval involves mismatching $S/N = E_b/N_0(1 - T_g/T)$, where $N_0/2$ is the Gaussian noise spectral density [1]. Here, for BPSK the guard time is set to be $T_g/T = 0.2$ leading to

$$T_g/T = 1/(4\log_2(M) + 1) \tag{9}$$

in general. It can be seen, that the loss of signal-to-noise ratio will be reduced for higher values M, e.g. $T_g/T = 1/21$ for $M = 32$.
For BPSK modulation, P_b is given by the well-known formula [6]

$$P_b = 0.5 \cdot \mathrm{erfc}\left(\sqrt{\gamma}\right). \tag{10}$$

Fig. 3 presents the analytical results of the error probabilities as a function of E_b/N_0. Different values M are considered for the Walsh modulation. The number of active users equals $J = 8$. In the case of BPSK and 2-ary Walsh modulation $N_2 = 64$ subcarriers are assumed. For the Walsh modulation it can

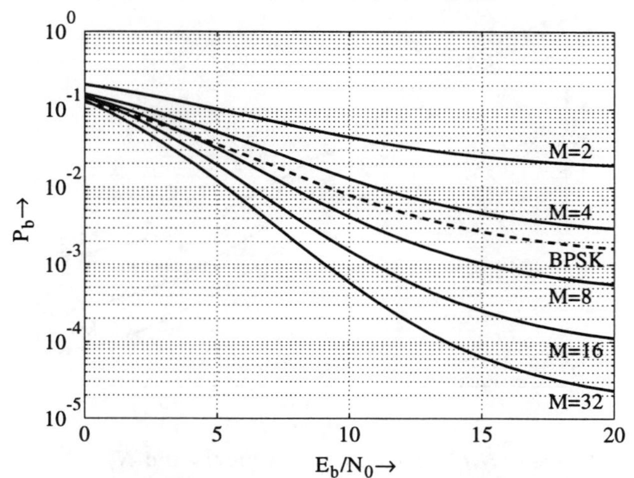

Figure 3: P_b versus E_b/N_0 for $J = 8$ active users and $N_2 = 64$ subcarriers

be seen that P_b decreases for larger values of M, which can be explained by the increasing Euclidian distance, see eq.(2). Since more subcarriers are used ($N_M = 64 \cdot \log_2(M)$), the MAI is also reduced. Walsh modulation outperforms BPSK for values $M \geq 8$ and $E_b/N_0 > 4$ dB. Modulation by $M = 32$ yields an error floor of $P_{b\infty} \approx 1.3 \cdot 10^{-5}$ instead of $P_{b\infty} \approx 1.2 \cdot 10^{-3}$ for BPSK.

4. INDOOR RAYLEIGH FADING CHANNEL

4.1. SYSTEM DESCRIPTION

The following results are given for an indoor Rayleigh fading channel. Assuming an exponential delay profile with a RMS delay of $\tau_{RMS} = 150\,\text{ns}$ and neglecting echos less than 10% of the maximum power yields a maximum delay spread of $\tau_{max} \approx 350\,\text{ns}$. The bandwidth is chosen to be $B = 25\,\text{MHz}$ in the 5.2 GHz range. A velocity of $v_0 = 0.5\,\text{m/s}$ results in a very low maximum Doppler frequency $f_{d,max}$ of about 9 Hz. Hence, long symbol durations T are possible. Assuming a constant guard time T_g, higher values of M therefore lead to less mismatching and better spectral efficiency. For example, if for BPSK $N_2 = 64$ subcarriers and $T_g/T = 0.2$ are considered ($T_g = 640\,\text{ns} > \tau_{max}$), mismatching will be reduced from 1 dB to 0.2 dB and the bit rate will be increased from 312 kbit/s to 372 kbit/s by raising M from 2 to 32.

4.2. SIMULATION RESULTS

Fig. 4 presents a performance comparison of BPSK and Walsh modulation. Assuming $J = 8$ active users and $N_2 = 64$ subcarriers, the bit error probability of Walsh modulation is shown for different values M. It can be seen that

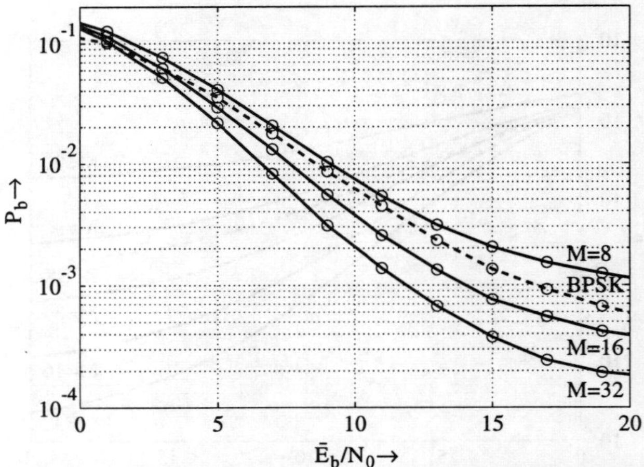

Figure 4: P_b versus $\bar{E}_b/N_0$ for $J = 8$ active users and $N_2 = 64$ subcarriers

Walsh modulation performs better than BPSK for $M \geq 16$ and an average $\bar{E}_b/N_0 > 4\,\text{dB}$. For the same number of active users, a bit error probability of $P_b = 10^{-3}$ is given for an 32-ary Walsh modulation $\bar{E}_b/N_0 = 12\,\text{dB}$ in contrast to a ratio of $\bar{E}_b/N_0 = 16.4\,\text{dB}$ for BPSK. Simulation of the error floors results in an approximately three times lower one for 32-ary Walsh modulation ($P_{b\infty} \approx 1.3 \cdot 10^{-4}$ for Walsh and $P_{b\infty} \approx 3.8 \cdot 10^{-4}$ for BPSK).

A comparison with the AWGN channel in fig. 3 illustrates a loss of the Walsh modulation performance. In the case of AWGN conditions, the Walsh modulation already performs better for $M \geq 8$. This effect can be explained by the deterioration of the orthogonality of the Walsh symbols. Since MRC is used for equalization, the orthogonality is not exactly restored which causes (self-)interference by all possible symbols applying the MLD.
To analyse the influence of the number of subcarriers, Tab. 1 presents the error probability for $N_2 = 64, 128$ subcarriers assuming $\bar{E}_b/N_0 = 12\,\text{dB}$. For

N_2	Walsh $M = 32$	BPSK
64	$P_b = 1.0 \cdot 10^{-3}$	$P_b = 3.1 \cdot 10^{-3}$
128	$P_b = 6.8 \cdot 10^{-5}$	$P_b = 8.0 \cdot 10^{-4}$

Table 1: P_b for $N_2 = 64, 128$ subcarriers with $\bar{E}_b/N_0 = 12$ dB and $J = 8$ active users

$N_2 = 64$, the 32-ary Walsh modulation performance is approximately three times and for $N_2 = 128$ it is even twelve times better. Hence, due to the fact of less MAI and less mismatching the improvement of M-ary Walsh modulation grows significantly. If we focus on the user interference of both schemes,

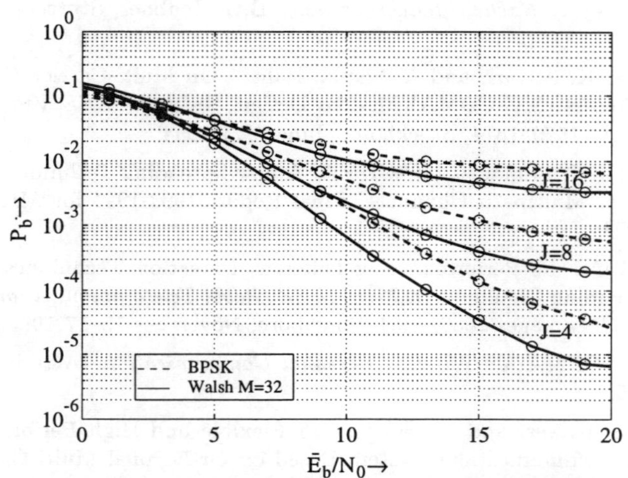

Figure 5: P_b versus $\bar{E}_b/N_0$ for $N_2 = 64$ subcarriers and different numbers of users

the results presented in fig. 5 show the tremendous MAI, which was expected for uplink conditions. Independent of the number of active users, the Walsh modulation performs better than BPSK.

5. CONCLUSION

In this paper, we introduced an M-ary orthogonal Walsh modulation concept for coherent reception under uplink conditions. In this case, we have an asynchronous reception and PN-sequences were applied. Performance evaluation has been presented for an AWGN channel. The results indicate that M-ary Walsh modulation significantly outperforms BPSK. Applying this modulation technique to an indoor Rayleigh fading channel further enhances the better performance. The concept of better using the available bandwidth by increasing the frequency resolution, combined with the advantages of raising Euclidian distance for larger values M, appears to be an interesting modulation scheme for mobile communication systems. Since the overall symbol duration grows for larger values M, which can be accepted for channels with low maximum Doppler frequency, mismatching is reduced and the spectral efficiency is increased. The results also indicate the MAI for traditional equalization such as MRC, which has been expected for uplink conditions. Therefore, uplink transmission with coherent detection requires new equalization schemes, e.g. multi-user detection. In addition, analyses have to be done for noncoherent transmission over uplink channels especially for indoor communication.

6. REFERENCES

[1] K.D. Kammeyer. *Nachrichtenübertragung*. B.G. Teubner, Stuttgart, second edition, 1996.

[2] G. Fettweis, K. Anvari, and A. Shaikh Bahai. On Multi-Carrier Code Division Multiple Access (MC-CDMA) Modem Design. In *44th IEEE Veh. Tech. Conf. (VTC)*, pages 1670–1674, Stockholm, June 7–11 1994.

[3] N. Yee and J.-P. Linnartz. MC-CDMA: A new Spreading Technique for Communication over Multipath Channels. Final Report 1993-1994 for MICRO Project 93-101, 1994.

[4] S. Kaiser. On the Performance of Different Detection Techniques for OFDM-CDMA in Fading Channels. In *Proc. IEEE Global Telecommunication Conference (GLOBECOM)*, pages 2059–2063, Singapore, November 13–17 1995.

[5] R. Prasad. *CDMA for Wireless Personal Communication*. Artech House, Norwood, 1996.

[6] K. Fazel, S. Kaiser, and M. Schnell. A Flexible and High Performance Cellular Mobile Communications System Based on Orthogonal Multi-Carrier SSMA. *Wireless Personal Communications*, 2:121–144, 1995.

[7] N. Yee, J.-P. Linnartz, and G. Fettweis. Multi-Carrier CDMA in Indoor Wireless Radio Networks. In *Proc. IEICE Transaction on Communications*, volume E77-B7, No.7, Japan, July 1994.

COMPARISON OF WLAN MULTICARRIER DS-SS PHYSICAL LAYER CONFIGURATIONS IN MEASURED INDOOR ENVIRONMENT

P. Banelli, G. Baruffa, S. Cacopardi, F. Frescura, G. Reali

Istituto di Elettronica, Università di Perugia -Via G. Duranti 1 - 06125 Perugia - Italy

Abstract

In this paper we compare two different MCM-DS-SS physical layer configurations for WLANs, by evaluating their performance in terms of bit error rate and frame error rate using measured channel delay profiles. Throughout a comparison of the obtained performance with the results of the DS-SS physical layer of the IEEE 802.11 standard, we point out some important architectural aspects and the potential limiting factors of the performance.

Introduction

Wireless local area networks (WLANs) represent an important field in the wireless communications industry, with a projected growth rate of 100% per year for the next three years. Some of the computer and communications industries' leading vendors are introducing Personal Digital Assistants (PDAs), modems, wireless interfaces and other devices and applications in support of wireless communications. WLANs offer new appealing features as they represent an alternative for adding new users to corporate LANs and supporting workers in remote locations, a low cost alternative to cable-base systems and a near ubiquitous possibility to access to any data base or any application located in the backbone. To satisfy the needs of wireless data networking, the study group 802.11 was formed under IEEE project 802 to recommend an international standard for WLANs [1]. Key parts of the standard are the Medium Access Control (MAC) protocols (to support asynchronous and time bounded delivery of data frames) and the DS-SS and FH-SS Physical Layer Specifications (PHY). Carrier Sense Multiple Access with Collision Avoidance (CSMA/CA) has been proposed as the basis for the IEEE 802.11 WLAN MAC protocols. The IEEE 802.11

K. Fazel and G.P. Fettweis (eds.), Multi-Carrier Spread-Spectrum, 77-86.

standard adopts the unlicensed ISM due to the available free bandwidth. Interference and multipath fading are the most significant issues in WLANs operating in the ISM band, where electronic devices radiate RF energy. In such an environment the interference rejection properties of the Spread Spectrum (SS) technique are precious [2]. Moreover, the adoption of the SS technique allows it to gain an intrinsic diversity of the signal in order to cope with the effects of multipath fading. In this paper we show the convenience of combining the SS signalling with the Multicarrier Modulation (MCM), to support indoor WLAN applications. The MCM was considered by the ETSI for the HIPERLAN standard [3], although Decision Feedback Equalisation (DFE) was finally adopted. The MCM approach, when combined with the SS technique, leads to some interesting technical solutions that exploit the strong issues of both techniques, such as frequency diversity, full channel equalisation by means of a Complex-valued Automatic Gain Control (CAGC) at the receiver side, and rejection of the interference [4][5]. An MCM-DS-SS radio subsystem for WLAN must be designed taking account of the special issues of such an application. Since a CSMA/CA access scheme is adopted, the Spread Spectrum technique is only devoted to the rejection of interference and to provide a proper diversity order [6]. Moreover, since a good value of throughput has to be guaranteed also for short packets and for handshake of MAC messages, a great care must be observed in the strategy adopted for the channel estimation and equalisation, by preferring simple techniques which allow to obtain small values of the processing delay.

II. Radio Subsystem Description

II.1 Modem Scheme

Figures 1 and 2 show the base-band schemes of the proposed MCM-DS-SS modems. The considered MCM-DS-SS schemes differ from each other in the strategy adopted for the sounding and equalisation of the channel. In the first MCM-DS-SS scheme (*System A*) the sounding of the channel is obtained by interleaving a number of known pilot carriers, frequency-spaced at least by one half the coherence bandwidth of the channel, in each MCM symbol. The receiver obtains an estimation of the channel frequency response by linearly interpolating the amplitude and phase values of the received pilot carriers. In the second MCM-DS-SS scheme (*System B*) the sounding of the channel is avoided by differentially encoding the complex-valued data symbols, before the IFFT processing, at the transmitter side. At the receiver side a dual operation is performed in order to reverse the differential encoding and to recover the effects of the channel. In both configurations, at the transmitter side the data frame is packed in blocks of μm bits. Each block is mapped into μ symbols of the M-PSK constellation (with $M=2^m$) according to the Gray code. The resulting complex-valued block is spread by μ periods of a unique spreading sequence (L being the period of the sequence). The spread block is serial to parallel converted in a vector of μL components and then scrambled by a block interleaver of depth equal to μ. In *system A* the sounding signal, which is formed of a number P_{pil} of known pilot carriers, is interleaved with the data symbols of each MCM block, just before the IFFT processing, while in *system B* a differential encoding of the data is performed by adding a reference symbol. In both cases, the resulting vector feeds an IFFT processor which operates on L_{fft} samples. In particular, in order to allocate the transition bandwidth of the transmitter filter, each vector is padded by

$N_1 = \lfloor (L_{fft} - (\mu L + P_{pil}))/2 \rfloor$ for *system A*, $N_1 = \lfloor (L_{fft} - (\mu L + 1))/2 \rfloor$ for *system B*, and $N_2 = L_{fft} - N_1$ zeros at its edges. To allow the operation in a frequency-selective fading environment, circular prefixes of duration T_g, larger than the delay spread of the channel, are added between adjacent IFFT transformed vectors in order to avoid inter-symbol interference (ISI) and inter-carriers interference (ICI). The resulting vector, after a final parallel to serial conversion, feeds the IF-RF sections.

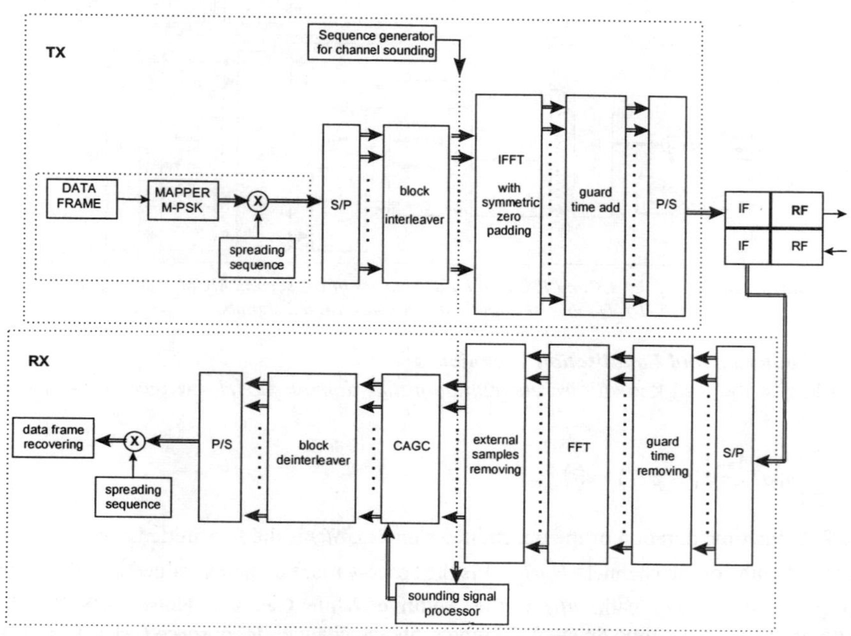

Fig. 1 Basic schemes of the TX and RX base-band sections of the first MCM-DS-SS modem (System A). Double lines indicate complex-valued signals.

The receiver sections of the modems in Figs. 1 and 2 are dual with respect to the transmitter. From the RF-IF sections the received signal is sampled and serial to parallel converted; then the circular prefix is removed. In *system A*, since the sounding signal is interleaved in each MCM symbol, the receiver performs an FFT processing on L_{fft} samples, eliminates the zeroes at the edges of the received vector to extract the $\mu L+P_{pil}$ central components, calculates the estimation of the channel (and consequently the CAGC coefficients) starting from the P_{pil} pilot carriers, and then processes the resulting vector (of μL components) by means of the CAGC and the deinterleaver. In *system B*, after the FFT processing, the differential decoding is performed to recover the effects of the channel selectivity, and then the resulting vector (of μL components) is deinterleaved. In both *systems A* and *B*, after a parallel to serial conversion, the signal feeds the spread spectrum receiver and then the data recovering circuits.

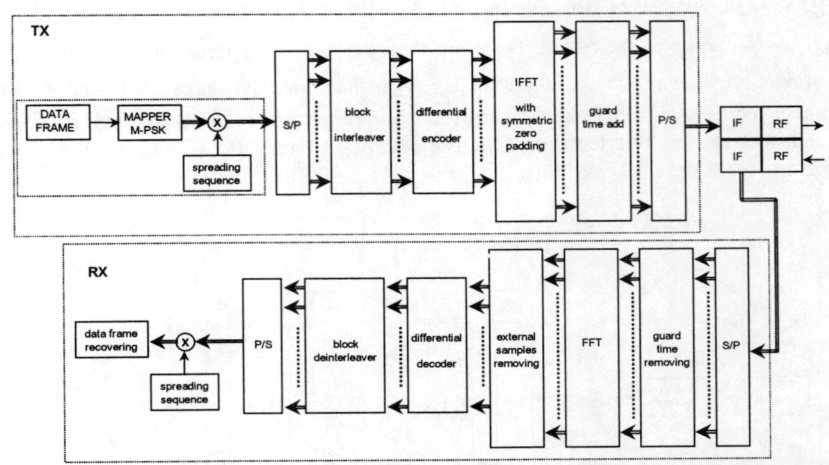

Fig. 2 *Basic schemes of the TX and RX base-band sections of the second MCM-DS-SS modem (system B). Double lines indicate complex-valued signals.*

II.2 Sounding and Equalisation Techniques

According to the well known *discrete multipath time-domain model*, the received signal can be written as

$$r(t)=\sum_{p=1}^{N_p} h(pT_S,s)S(t-pT_S)+n(t) \tag{1}$$

were T_S is the time duration of the transmitted samples, $S(t)$ is the transmitted signal, N_p is the number of paths of the channel, $h(pT_s,s)$ is the space-variant complex-valued coefficient that corresponds to the pth path, $n(t)$ is the Additive White Gaussian Noise (AWGN). The maximum value of the time delay, N_pT_S represents the channel *delay spread.* Because of the small random variations of environment and user mobility, the space-variant nature of the channel translates itself in a time dependence of the channel impulse response. If the duration $T_b=L_{fft}T_S$ of transmitted OFDM symbols (L_{fft} components) is considerably smaller than the coherence time ($\Delta T_c=1/f_D$, f_D being the maximum Doppler frequency) of the channel, the slow fading channel condition is maintained, thus expression (1) can be written as

$$\mathbf{R}=\mathbf{h}*\mathbf{S}_{ifft}+\mathbf{N} \tag{2}$$

In (2) the symbol "*" denotes the linear discrete convolution, $\mathbf{S}_{ifft}=[s_1,s_2,...,s_{L_{fft}}]$ and $\mathbf{R}=[r_1,r_2,...,r_{L_{fft}+N_p-1}]$ are the transmitted and received vectors respectively, $\mathbf{N}=[n_1,n_2,...,n_{L_{fft}+N_p-1}]$ and $\mathbf{h}=[h_1,h_2,..h_{N_p}]=[h(T_S,s),h(2T_S,s)..h(N_pT_S,s)]$ are the AWGN vector and the discrete channel impulse response at the sounding time respectively. In the following a description of the sounding and equalisation techniques of the two considered systems will be provided.

System A: In this case the sounding signal is interleaved with data, in each MCM symbol, before the IFFT processing at the transmitter, according to a technique similar to the one adopted in the Digital Video Broadcasting (DVB-T) standard [7]. The sounding vector is defined as

$$S_{pil}[k] = \sqrt{\frac{\gamma_P}{P_{pil}}} \sum_{n=0}^{P_{pil}-1} F_{SC}[k]\delta(k - n(L+1)), \; 0 \le k < \mu L + P_{pil}, \tag{3}$$

where P_{pil} is the number of pilot carriers in each MCM block, γ_p is the energy value of the sounding vector and $\mathbf{F}_{SC}$, with $|F_{SC}[k]| = 1$, is a random complex valued vector which is adopted to scatter the pilot carriers in order to reduce the crest-factor (the ratio between the maximum and the mean value of the magnitude) of the transmitted signal. Starting from the data vector $\mathbf{d} = [d_1, d_2, ..., d_{\mu L}]$, the following is obtained

$$\mathbf{D}_{pil} = [0, d_1, d_2, .., d_L, 0, d_{L+1}, d_{L+2}, .., d_{2L}, 0, ., d_{(\mu-1)L+1}, d_{(\mu-1)L+2}, .., d_{\mu L}, 0], \tag{4}$$

by interleaving P_{pil} zeroes in the positions corresponding to the pilot carriers in (3). After the combination of sounding and data signals

$$\mathbf{S} = \mathbf{S}_{pil} + \mathbf{D}_{pil}, \tag{5}$$

the resulting vector **S** is IFFT processed and transmitted. At the receiver side, after the FFT processing on L_{fft} samples and the elimination of the virtual carriers to extract the $\mu L + P_{pil}$ central components, the estimation of the channel is obtained by extracting the P_{pil} pilot carriers (in the following the symbol "$\otimes$" and "$\div$" denote a complex multiplication and a division component by component, respectively):

$$\hat{\mathbf{D}}_{pil} = \mathbf{R}_{fft} \otimes \mathbf{M} \div \mathbf{F}_{SC}, \; \mathbf{M} = \overbrace{[1, \underbrace{0,0,..0}_{L}, 1, \underbrace{0,0,..0}_{L}, 1, .., \underbrace{0,0,..,0}_{L}, 1]}^{\mu L + P_{pil}}, \tag{6}$$

$\mathbf{R}_{fft}$ being the received vector (2) after the FFT processing. The resulting vector (6) is then interpolated to provide $\hat{\mathbf{H}} = [\hat{h}_{N_1+1}, \hat{h}_{N_1+2}, ..\hat{h}_{L_{fft}-N_2}]$, which represents the estimation of the channel transfer function. An improvement of the quality of the estimation of the channel is obtained by filtering the $\hat{\mathbf{H}}$ by means of a *zero-phase forward and reverse digital filter* [8], whose pass-band is determined according to the frequency spacing of the pilot carriers and to the coherence bandwidth of the channel. From the estimated channel response $\hat{\mathbf{H}}$, the CAGC coefficients are obtained as the conjugate of each component, according to the *Maximum Ratio Combining* equalisation technique:

$$C[k] = \tilde{H}^*[k], \; k = N_1 + 1, ..., L_{fft} - N_2. \tag{7}$$

System B: The estimation and the equalisation of the channel can be avoided if a proper differential encoding of the data carriers is performed. A differential scheme which involves the corresponding carriers of consecutive MCM symbols is adopted in the Digital Audio Broadcasting (DAB) standard [9]. Differently, in *system B*, in order to reduce the overhead due to the transmission of a dedicated reference MCM block at the beginning of each data packet, a differential encoding of adjacent carriers in each MCM symbol has been selected. Starting from the data vector $\mathbf{d} = [d_1, d_2, ..., d_k d_{k+1}, ..., d_{\mu L}]$, where the indexes k and $k+1$ refer to two adjacent carriers before the deinterleaver, the following quantities may be defined:

k-th M-PSK symbol: $d_k = \rho_k e^{j\theta_k}$ (8)

k-th M-DPSK encoded symbol: $s_k = \sigma_k e^{j\varphi_k}$ (9)

(k+1)-th M-PSK symbol: $d_{k+1} = \rho_{k+1} e^{j\theta_{k+1}}$ (10)

(k+1)-th M-DPSK encoded symbol: $s_{k+1} = \sigma_{k+1} e^{j\varphi_{k+1}} = s_k d_{k+1} = \sigma_k \rho_{k+1} e^{j(\varphi_k + \theta_{k+1})}$ (11)

At the receiver side, after the FFT processing, the corresponding symbols are:

k-th M-DPSK received symbol: $r_k = s_k h_k + n_k$ (12)

(k+1)-th M-DPSK received symbol: $r_{k+1} = s_{k+1} h_{k+1} + n_{k+1}$, (13)

the complex-valued coefficients, h_k and h_{k+1} being the components of the channel frequency response and n_k e n_{k+1} the AWGN samples. The differential decoder processes the symbols as follows:

(k+1)-th M-PSK decoded symbol: $\hat{d}_{k+1} = r_{k+1} r_k^* = (s_{k+1} h_{k+1} + n_{k+1})(s_k^* h_k^* + n_k^*)$. (14)

By neglecting the variation of the channel frequency response between adjacent carriers (i.e. considering $h_k = h_{k+1}$), and taking into account that $\sigma_k = \theta_k = 1, \forall k$, relation 14 can be rewritten as

$$\begin{aligned}\hat{d}_{k+1} = r_{k+1} r_k^* &= \left(s_{k+1} s_k^* h_{k+1} h_k^* + s_{k+1} h_{k+1} n_k^* + s_k^* h_k^* n_{k+1} + n_{k+1} n_k^*\right) = \\ &= |h_k|^2 \left(\sigma_k \rho_{k+1} e^{j(\varphi_k + \theta_{k+1})}\right)\left(\sigma_k e^{-j\varphi_k}\right) + s_{k+1} h_{k+1} n_k^* + s_k^* h_k^* n_{k+1} + n_{k+1} n_k^* = \\ &= |h_k|^2 e^{-j\theta_{k+1}} + s_{k+1} h_{k+1} n_k^* + s_k^* h_k^* n_{k+1} + n_{k+1} n_k^*.\end{aligned} \tag{15}$$

Expression 15 shows that, apart from the noise terms, the combined effect of (11) and (14) is to recover the phase shift due to channel and to produce a weighting on each carrier which is identical to that produced by the *Maximum Ratio Combining* equalisation technique, adopted in *system A*. It should be noted that the absence of channel sounding leads to a simplified hardware structure and to reduced values of processing delays.

II.3 System and Frame Parameters

Table 1 shows the most significant system parameters of the considered configurations. In order to make a number of considerations and comparisons possible, we will assume as a reference some of the system and frame parameters of the IEEE 802.11 Draft Standard [1]. From Table 1 the following issues should be pointed out:

- The proposed system configurations are characterised by the same *chip rate* and the same *spreading factor L* of the IEEE 802.11 DS-SS Physical Layer Specifications [1].
- The measured maximum *delay spread* is 1 μs, that leads to a *coherence bandwidth* of 1 MHz. By comparing these values with the *spreading factor L* and the *channel bandwidth* it

is possible to conclude that a full exploitation of the diversity order of the channel is achieved.

- The *FFT length* is 512 samples for both *system A* and *system B* configurations. This value represents a good compromise among some aspects such as: efficiency with respect to the adoption of a circular prefix between adjacent MCM blocks, system throughput for short packets and for MAC message handshake, implementation of a real-time FFT / IFFT processor.

MCM-DS-SS WLAN SYSTEMS PARAMETERS			
MAX DELAY SPREAD (s)	1,00E-06	CARRIER FREQUENCY (Hz)	2,40E+09
COHERENCE BANDWIDTH (Hz)	1,00E+06	MAX SPEED RX-TX (m/s)	1,00
CHANNEL BANDWIDTH (Hz)	1,10E+07	MAX DOPPLER SPREAD (Hz)	8,00
CHIP RATE (chip/s)	1,10E+07	CHANNEL COHERENCE TIME (s)	1,25E-01
CHIP DURATION (s)	9,09E-08		
CIRCULAR PREFIX LENGTH (chips)	11		

System A		System B	
FRAME MAX LENGTH (bytes)	2400	FRAME MAX LENGTH (bytes)	2400
FFT LENGTH (samples)	512	FFT LENGTH (samples)	512
SPREADING FACTOR	11	SPREADING FACTOR	11
DATA SYMBOLS / MCM SYMBOL	40	DATA SYMBOLS / MCM SYMBOL	43
DATA SYMBOLS MAPPING	8-PSK	DATA SYMBOLS MAPPING	8-PSK
BITS / MCM SYMBOL	120	BITS / MCM SYMBOL	129
BYTES / MCM SYMBOL	15	BYTES / MCM SYMBOL	16
EFFECTIVE BITS / MCM SYMBOL	120	EFFECTIVE BITS / MCM SYMBOL	128
MCM DATA SYMBOLS / FRAME	160	MCM DATA SYMBOLS / FRAME	150
DATA CARRIERS	440	DATA CARRIERS	474
PILOT CARRIERS	41	PILOT CARRIERS	0
VIRTUAL CARRIERS	31	VIRTUAL CARRIERS	38
MCM CARRIERS SPACING (Hz)	21484,4	MCM CARRIERS SPACING (Hz)	21484,4
PILOT CARRIERS SPACING (Hz)	252048,4	PILOT CARRIERS SPACING (Hz)	-----
GUARD BAND (kHz)	666,0	GUARD BAND (kHz)	816,4
MCM SYMBOL RATE (symbol/s)	21032	MCM SYMBOL RATE (symbol/s)	21032
DATA BLOCK RATE (blocks/s)	21032	DATA BLOCK RATE (blocks/s)	21032
MCM SYMBOL DURATION (s)	4,75E-05	MCM SYMBOL DURATION (s)	4,75E-05
CHANNEL ESTIMATION PERIOD (s)	4,75E-05	CHANNEL ESTIMATION PERIOD (s)	4,75E-05
BIT RATE (bit/sec)	2.523.840	BIT RATE (bit/sec)	2.692.096
SYMBOL RATE	841.280	SYMBOL RATE	897.365

Table 1 *System parameters of the OM-DS-SS WLAN system.*

- The number of 8-PSK *data symbols* in each MCM block is 40 and 43 for *system A* and *system B* respectively. This fact leads to slightly different values of spectral efficiency with *system B* taking advantage by the absence of pilot carriers in each block.
- The comparison of the *channel estimation period* of both configurations with the *channel coherence time* and the comparison of the *pilot carriers spacing* of the *system A* with the channel *coherence bandwidth* lead to the expectation of a good behaviour of both systems with respect to the estimation and tracking of the channel time-variant response.
- The maximum achievable *bit rate* is 2,523,840 bit/s for *system A* and 2,692,096 bit/s for *system B*. Taking into account the value of the *channel bandwidth* of 11 MHz it can be argued that the proposed system configurations achieve a maximum spectral efficiency value more than doubled with respect to the IEEE 802.11 (2 Mbit/s in one channel of 22 MHz).

Fig. 3 shows the frame structure of the proposed MCM-DS-SS physical layer configurations. In *system A* a frame of 2400 bytes is segmented into 160 MCM symbols. The sounding signal is embedded in each MCM block containing 40, 8-PSK symbols, corresponding to 15 bytes. Since no dedicated sounding block is required, the shortest allowed block is represented by a single 15 bytes data block. In the *system B* configuration a frame of 2400 bytes is segmented into 150 MCM symbols. The shortest allowed block is represented by a single 16 bytes data block.

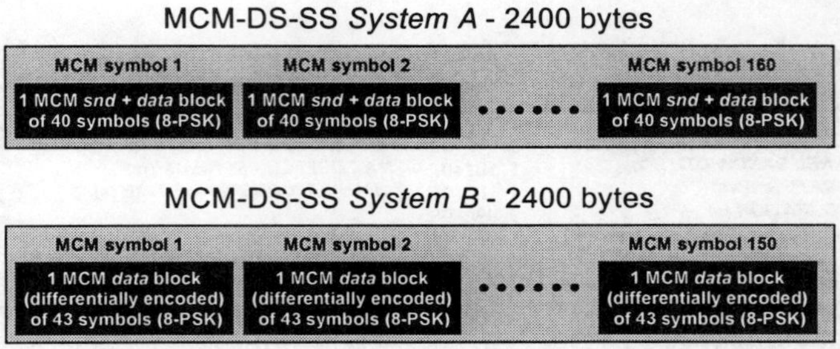

MCM-DS-SS *System A* - 2400 bytes

MCM symbol 1: 1 MCM *snd* + *data* block of 40 symbols (8-PSK)
MCM symbol 2: 1 MCM *snd* + *data* block of 40 symbols (8-PSK)
......
MCM symbol 160: 1 MCM *snd* + *data* block of 40 symbols (8-PSK)

MCM-DS-SS *System B* - 2400 bytes

MCM symbol 1: 1 MCM *data* block (differentially encoded) of 43 symbols (8-PSK)
MCM symbol 2: 1 MCM *data* block (differentially encoded) of 43 symbols (8-PSK)
......
MCM symbol 150: 1 MCM *data* block (differentially encoded) of 43 symbols (8-PSK)

Fig. 3 *Frame format of the MCM-DS-SS radio access schemes.*

V. Numerical Results

A set of simulation trials has been carried out in order to point out the following aspects:

- *crest factor*: this key parameter (defined as the ratio between the maximum and the mean values of the transmitted signal magnitude) has a great impact on the implementation issues of MCM systems, since it gives a measure of the dynamic range required by the power amplifier of the TX section of the modem.
- *BER and FER performance*: Simulation trials in LOS and NLOS conditions have been carried out to verify the performance of the proposed configurations throughout a comparison with the IEEE 802.11 DS-SS Physical Layer.

Figure 4 shows the crest factor of both *system A* (with or without the scattering of pilot carriers) for different values of the S/D ratio (defined as the energy ratio between the sounding and data signals in each MCM symbol) and *system B*. Without the scattering of the pilot carriers *system A* shows a crest factor which decreases, as expected, with the decreasing of the S/D ratio. The minimum value of 8.2 dB can be considered as driven by the data signal. The carrier scattering produces a significant reduction. The low value of crest factor for S/D=0dB may look strange, but it is explained by the fact that the scattering produces nearly constant signal peak values, while the signal mean values depend directly on the S/D ratio. *System B* shows a constant value of 8.2 dB, since the crest factor is driven only by the data signal.

Figure 5 and 6 show the BER and FER versus data *Eb/No* of *systems A* and *B*, in LOS and NLOS conditions respectively, compared with the performance of IEEE 802.11 DS-SS physical layer. The frame length is half the maximum frame length for the three systems, while the S/D ratio is equal to -3dB for *system A* configuration. The two MCM systems show a performance practically insensitive with respect to the channel, both in terms of BER and FER. The performance penalty of *system B* with respect to *system A* is about 2.7 dB, but if the total transmitted power (data + sounding) is considered, the corresponding performance penalty is less than 1dB.

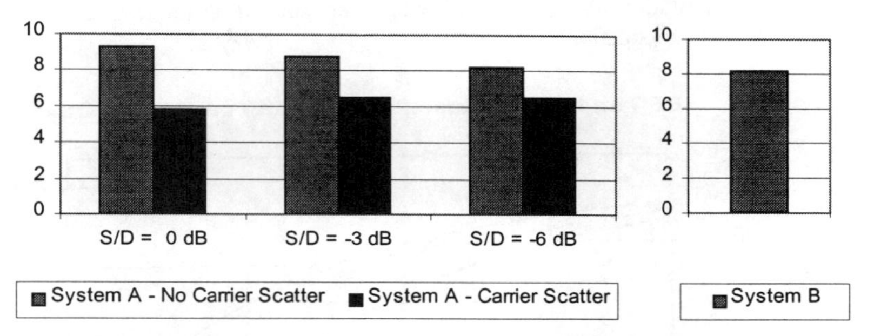

Fig. 4 *Crest factor of the MCM-DS-SS radio access schemes for different values of the S/D ratio.*

The IEEE 802.11 DS-SS, with DQPSK modulation, shows a performance comparable to the best of the MCM systems only for low values of *Eb/No*. When the quality of the channel equalisation becomes significant, i.e. high values of *Eb/No* and/or NLOS conditions, the performance margin of both the MCM schemes becomes evident.

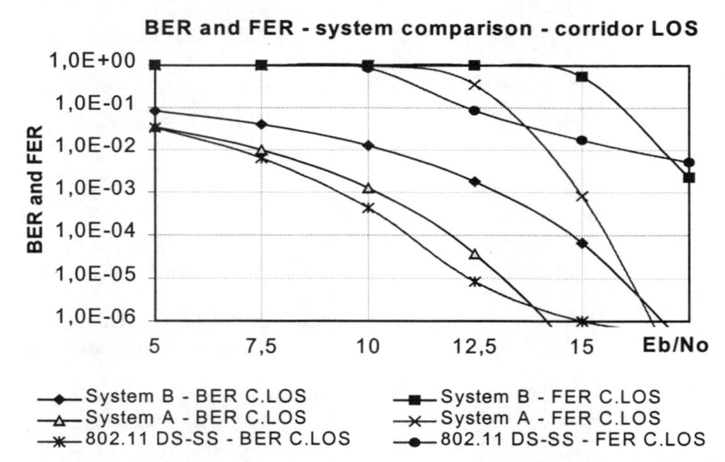

Fig. 5 *BER and FER simulated performance of systems A and B in LOS conditions compared with the performance of IEEE 802.11 DS-SS physical layer.*

Conclusions

In this paper we have proposed two different MCM-DS-SS physical layer configurations for WLANs. The systems analysed show a spectral efficiency value more than doubled with respect to the IEEE 802.11 and show a good performance in terms of BER thanks to a good capacity for counteracting multipath fading. In particular the differential configuration (*System B*) shows an interesting performance in terms of power efficiency and a significantly simplified hardware configuration, since the estimation of the channel is avoided. As a conclusion, if technology can provide efficient and low cost solutions for modem chipsets, the differential configuration of MCM-DS-SS can be considered an interesting technique to be adopted for WLAN applications.

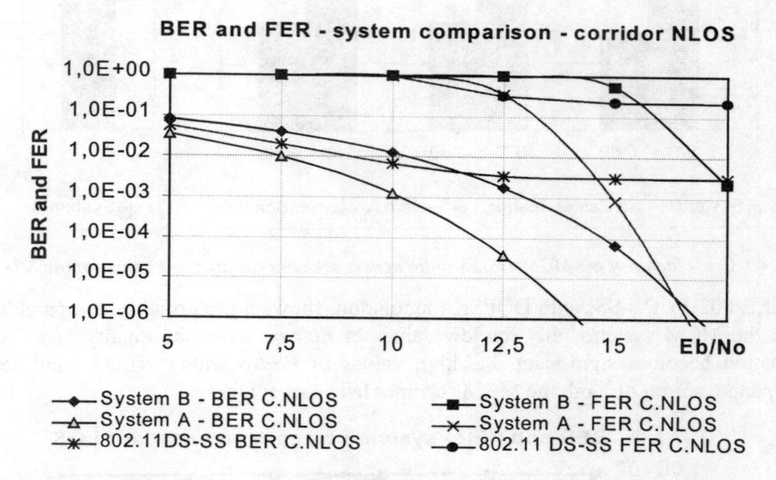

Fig. 6 *BER and FER simulated performance of systems A and B in NLOS conditions compared with the performance of IEEE 802.11 DS-SS physical layer.*

References

[1] IEEE 802.11/D1, "Wireless LAN Medium Access Control (MAC) and Physical Layer (PHY) Specifications", Draft Standard, 1 December 1994.

[2] A. Falsafi, K. Pahlavan, "A Comparison Between the Performance of FHSS and DSSS for Wireless LANs Using a 3D Ray Tracing Program", IEEE Int. Conf. on Personal, Indoor and Mobile Radio Comm. (PIMRC'95), Toronto, Canada, pp 569-573, Sept. 1995.

[3] B. Bourin, "HIPERLAN - Markets and Applications Standardisations Issues", IEEE Int. Conf. on Personal, Indoor and Mobile Radio Comm. (PIMRC'94), Hague, Netherlands, pp 863-868, Sept. 1994.

[4] A. Chouly, A. Brajal, S, Jourdan, "Orthogonal Multicarrier Technique applied to Direct Sequence Spread Spectrum CDMA systems" IEEE GlobeCom 1993, Conference Record, pp. 1723-1728.

[5] N. Yee, J.P. M.G. Linnartz, G. Fettweis, "Multi-Carrier CDMA in Indoor Wireless Radio Networks", IEICE Trans. on Comm., vol. E77-B, July 1994.

[6] M. Schnell, S. Kaiser, "Diversity Consideration for MC-CDMA systems in Mobile Communications" IEEE International Symposium on Spread Spectrum Techniques and Applications (ISSSTA'96) - Mainz, Germany, September 1996.

[7] DVB, "Framing Structure, Channel Coding and Modulation for Digital Terrestrial Television" *DVB Document A012*, June 1996.

[8] A. V. Oppenheim, R. W. Schafer, "Digital Signal Processing", *Prentice-Hall*, 1975.

[9] ITU-R Special Publication on Digital Sound Broadcasting, DRAFT Version 2.0.

Section III

CODING AND DETECTION FOR MC-SS

MULTI-CARRIER CDMA USING CONVOLUTIONAL CODING AND INTERFERENCE CANCELLATION OVER FADING CHANNELS

J J Maxey and R F Ormondroyd

School of Electrical Engineering, The University of Bath, Bath, BA2 7AY, U.K
E-mail: eepjjm@bath.ac.uk and eesrfo@bath.ac.uk

ABSTRACT

OFDM DS-CDMA systems using convolutional coding techniques and multi-access interference (MAI) cancellation can be implemented effectively on the uplink of a mobile radio channel as long as the interference statistics are known at the receiver. The choice of the right equalisation technique is vital to maximise the SNR at the input to the detector. The use of maximum ratio combining (MRC) after MAI cancellation has been found to give good results. In this paper we compare the performance of MAI cancellation using zero-forcing equalisers and MRC with the performance of MAI cancellation using controlled equalisation paired with MRC. It is found that there are considerable performance improvements when controlled equalisation techniques are used for the initial estimate of the interfering user's data.

INTRODUCTION

Multi-Carrier OFDM techniques are increasingly gaining a wide interest in the field of mobile and fixed wireless communication systems using multiple access techniques such as DS-CDMA spread spectrum. Recently, a few papers have proposed the use of OFDM DS-CDMA for the uplink [1] in frequency selective faded channels with high interference. OFDM DS-CDMA can provide resilience to frequency selective faded channels by transforming high bit-rate data to multiple low-rate data streams, multiplexed over a wider bandwidth than the original data. This has the advantage of enabling the received data streams to appear as flat faded

K. Fazel and G.P. Fettweis (eds.), Multi-Carrier Spread-Spectrum, 89-96.

data streams on each sub-carrier, and this enables the equaliser in the receiver to demodulate the data more effectively. A number of techniques have been proposed to combat these fading effects:

- Frequency- or time domain interleaving
- Convolutional coding architectures
- Guard-time interval insertion
- Equalisation structures
- MAI cancellation techniques (successive or parallel)

If the time-delay spread on the channel is large compared to the chip period it becomes necessary to make use of guard-intervals to reduce the effects of inter-symbol interference (ISI). No ISI will occur as long as the guard-interval length is greater than the maximum delay spread of the channel impulse response. Using DS-CDMA to spread the original bandwidth of the data enables the delay spread of the channel to be resolved more efficiently. Therefore DS-CDMA techniques can provide a robust multiple access scheme when many other users are transmitting over the same bandwidth in a faded channel.

Although the time guard-band limits the effect of ISI, there is still a need for channel equalisation. OFDM allows equalisation to be performed in the frequency domain at the output of the receiver FFT and this allows a simple one tap equalisation structure to be used, with its channel knowledge derived from the FFT of the channel impulse response. This can be measured using pilot channels or midambles inserted between the data. A number of techniques using equalisers of different complexity have been investigated [e.g. 2,3]. When the receiver has no knowledge of either the received SNR or the number of users using the channel, a relatively simple and effective equaliser that attempts to restore the orthogonality of the sub-channels is *controlled equalisation*. For more sophisticated receiver structures, the MMSE algorithm can provide considerable performance improvements. Sub-optimal MMSE algorithms can be employed through fixed estimates of the number of users present in the channel. This presents about 1dB loss in performance for the downlink channel.

For reception of the wanted data in other-user noise, maximum likelihood detection schemes give optimum performance. However, this can only be used if the spreading sequences of all users are relatively short since the complexity increases exponentially to the sequence length. The use of convolutional coding on the data provides BER performance benefits in memoryless and non-memoryless channels, yielding a coding gain larger than the bandwidth penalty. In Rayleigh faded channels, the coding gain can be further enhanced through strong interleaving in the time or frequency domain. For multi-user systems, the interference is dominated by "other-user" noise and this becomes the limiting factor on the performance. A variety of interference cancellation schemes can be used to remove this interference, as long as the spreading codes of all active users are known at the receiver.

This paper presents a comparison of different equalisation strategies used in conjunction with MAI cancellation. In particular, we consider zero-forcing equalisation, controlled equalisation and maximum ratio combing (MRC).

UPLINK DESIGN

There have been a number of papers reporting efficient uplink strategies for a cellular mobile radio system. The uplink design investigated here is based on a service provision of 32 kb/s for each user, using OFDM modulation and low-rate orthogonal convolutional (LROC) coding to provide the necessary spreading and coding. It is assumed that a small overhead is needed to provide the basestation with an estimate of the channel transfer function of each user. This allows the basestation to equalise each user's data more effectively. A diagram of the proposed transmitter and receiver is shown in figures 1 and 2. Here it assumed that the wanted user is user 1, and the interference effects of users 2 to L (where L is the total number of simultaneous users) are partially removed through interference cancellation.

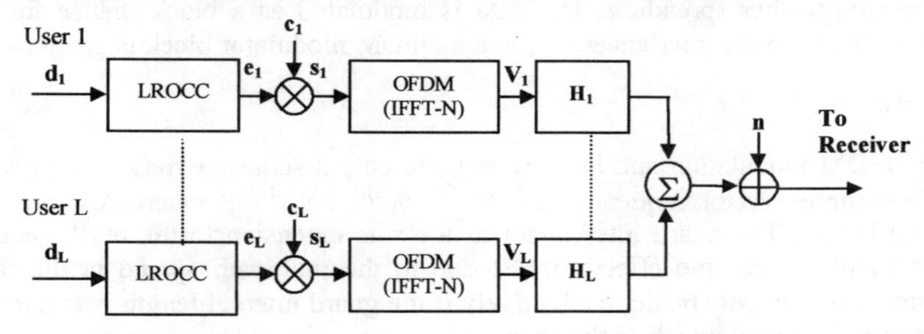

Figure 1. Simulation model of transmitter design for all uplink users and the channel model

Successive interference cancellation schemes are popular for designs that use no power control and perform best when all users in the system have geometrically distributed received powers. Parallel interference cancellation schemes, on the other hand, perform better when all users are assumed to be perfectly power controlled.

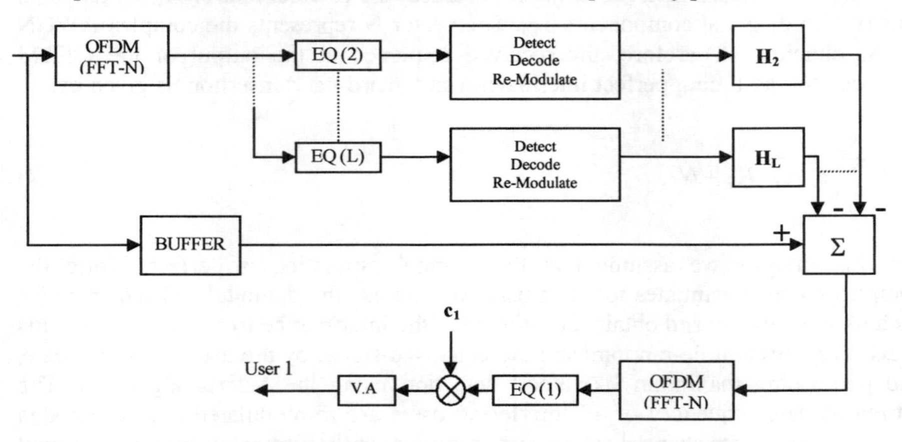

Figure 2. Basestation receiver structure to detect user 1

In the analysis we ignore the time index and consider the processing of a single block of data bits from user i. Each user transmits a data sequence $\boldsymbol{d_i}$, where i = 1,2 ... L. Each data bit is of duration T_b seconds and is consequently coded in a convolutional encoder of rate $R=K/M$. The input sequence $\boldsymbol{d_i}$ of K data bits produces an output sequence $\boldsymbol{e_i}$ of length n = [1 ... M], where each subsequent output symbol $e_i(n)$ is of duration $T_c=RT_b$. For LROCC codes the code rate is given as $R=1/M$, where M defines the spreading ratio of the user specific output sequence and is given as $M=2^K$. This equates to the bandwidth expansion factor. The encoded sequence consisting of orthogonal Walsh-Hadamard codes is further multiplied by a randomising sequence of equal length M, given as $\boldsymbol{c_i} = [c_i^{(1)}, c_i^{(2)} \ldots c_i^{(M)}]^T$, where $[.]^T$ denotes the transposition of the matrix. Note that the randomising sequence $\boldsymbol{c_i}$ does not provide further spreading. The data is modulated as a block, hence for each block the input to the interleaver within the OFDM modulator block is given as:

$$\boldsymbol{s_i}^{(n)} = \boldsymbol{e_i}^{(n)} \boldsymbol{c_i}^{(n)} \quad \textbf{(1)}$$

After OFDM modulation and interleaving, the output sequence may be represented by the complex vector sequence $\boldsymbol{V_i} = [v_i^{(1)}, v_i^{(2)} \ldots v_i^{(M+\Delta)}]^T$ where Δ is the guard interval length. The guard interval forms a cyclic extension to the block sequence [1...M] and reduces the effects of ISI due to the multipath spread in the fading channel. This can only be done effectively if the guard interval length is greater than the multipath spread length of the channel.

The output vector to be transmitted over the channel is represented by complex coefficients at the output of the IFFT in the modulator. Since each user communicating to the basestation experiences different independent fading statistics, the channel coefficients are different for each user. The channel may be described by a complex matrix $\mathbf{H_i}$ (in the frequency domain) of size $M \times M$ for each user, affecting the subcarriers $\mathbf{V_i}$ assigned to the transmitted sequence on the transmitter for user i. Since a perfect guard-interval length is assumed, the channel matrix $\mathbf{H_i}$ is a diagonal matrix with diagonal components $h_{n,n}$. The vector $\mathbf{N}$ represents the complex AWGN in the channel. Therefore, the received sequence at the output of the OFDM demodulator, assuming perfect interleaving and guard-band insertion, is given by:

$$\boldsymbol{R} = \sum_{i}^{N_u} \boldsymbol{e_i} \cdot \boldsymbol{c_i} \cdot \boldsymbol{H_i} + \boldsymbol{N} \quad \textbf{(2)}$$

For this analysis we assume that the channel estimation is perfect. Using the complex channel estimates for each user we equalise the demodulated sequence for each user separately and obtain an estimate of the interference from other users. This is achieved through de-randomising the equalised signal by the user specific code $\boldsymbol{c_i}$ and performing maximum likelihood detection using the Viterbi algorithm. The estimated data sequences of all interfering users are re-modulated and re-encoded with the appropriate channel estimates and subsequently subtracted from the original received sequence. Depending on the level of error in the estimation of the interfering user's data sequences, the bit-error performance of the wanted user can vary drastically.

The initial detection involves equalising structures that try to restore the shape of the signals in order to detect the other users' interference. The first stage of detection uses either zero forcing or controlled equalisation methods, as discussed in [4]. For zero-forcing equalisation, the effective coefficients are given as:

$$g_{n,n} = \frac{1}{h_{n,n}} = \frac{h_{n,n}{}^*}{\left|h_{n,n}\right|^2} \tag{3}$$

If controlled equalisation is used in the receiver, a rapid deep fade in the channel transfer function below a set threshold value will cause the receiver to switch to phase only equalisation on that particular carrier. The complex equalisation coefficient is then given by:

$$g_{n,n} = \frac{h_{n,n}{}^*}{\left|h_{n,n}\right|} \tag{4}$$

These enable the receiver to obtain an initial estimate of the interfering user's data sequences. When the interfering signals have been eliminated from the original received signal, it is then important to maximise the signal-to-noise ratio in the detector. For this reason it is of interest to use a combiner such as MRC. The complex equalisation coefficient is given in the frequency domain as:

$$g_{n,n} = h_{n,n}{}^* \tag{5}$$

MRC provides a signal estimate by combing the maximum signal-to-noise ratio of each faded sub-carrier and performs best on "clean" signals. Therefore this combining method is used on the second iteration. This provides an optimal equalisation strategy, since the "cleaned" signals are assumed to be relatively free of other-user interference.

SIMULATION RESULTS

The simulated uplink channel assumes a service provision of a 32 kb/s data stream for each user. A constraint length of K=5 provides a spreading ratio of 32 in the LROCC encoder and this is then randomised through the user-specific Gold code c_i of the same rate. This code provides no further additional spreading, but merely serves as a randomising sequence for the LROCC orthogonal code sets [5,6]. The number of sub-carriers is N=32 in a bandwidth of 1.028 MHz, and therefore each sub-carrier has a duration of 31.25μs. The simulated channel is based on the COST 207 frequency selective Rayleigh faded bad urban (BU) channel model [7]. The

Doppler frequency was set to 200Hz and perfect power control is assumed. Figure 3 shows the BER performance for different numbers of simultaneous users communicating with the basestation. The performance without MAI cancellation degrades significantly when more than 5 users are present. The use of MAI cancellation increases this to at least 10 users, and this is a significant gain in

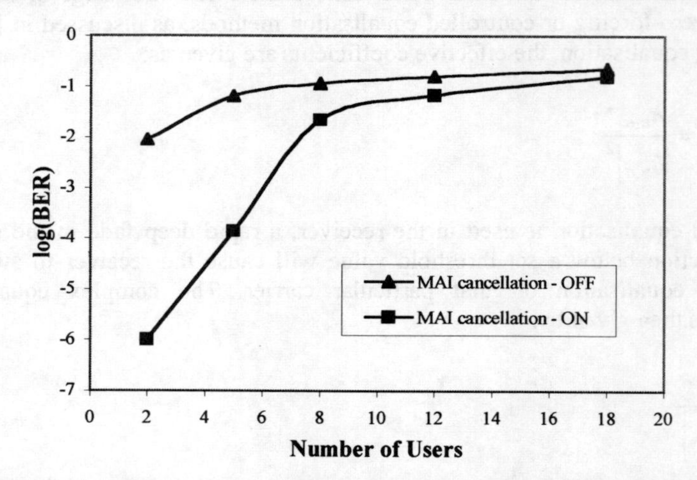

Figure 4. Uplink capacity comparison using zero forcing equalisation on the first iteration and MRC on the second detection stage

performance. Conversely, at low user numbers, the BER for 32 kb/s per user is substantially improved by MAI cancellation. Without MAI cancellation it is not possible to achieve an acceptable BER for any user numbers.

Controlled threshold equalisation tries to balance the disadvantages of zero forcing and phase only equalisation in fading conditions to provide an optimum balance of equalisation. The impact of threshold level on BER is shown in figure 4 using no MAI cancellation and 8 simultaneous users in the channel.

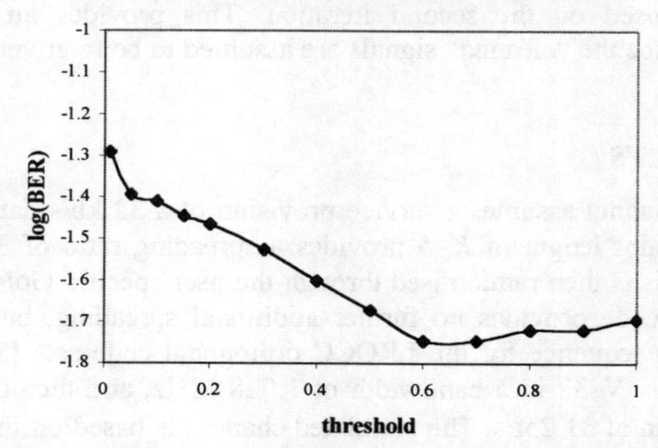

Figure 3. Optimum threshold value for controlled equalisation on initial detection

This figure suggests an optimum threshold value of about 0.6, and this value was subsequently used for the second type of receiver system that has been simulated which uses a controlled equaliser. The performance of this receiver is shown in figure 5.

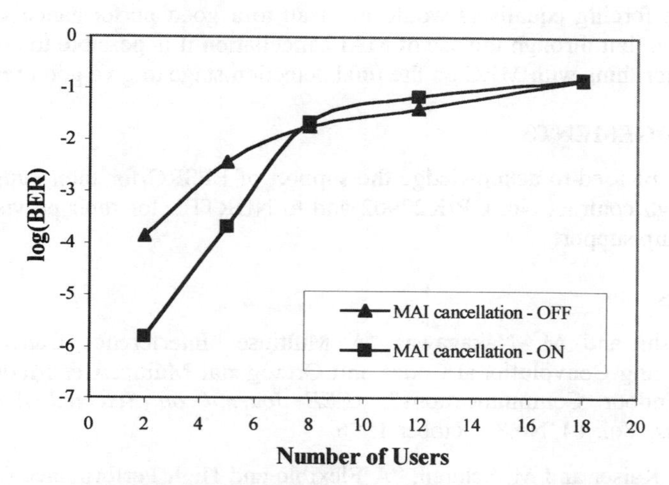

Figure 5. Uplink capacity comparison using controlled equalisation on the first iteration and MRC on the second detection stage

Compared with the method that uses the zero-forcing strategy on the initial iteration, described earlier, this technique shows a significant performance improvement when no MAI cancellation is used in other user noise environments, but a smaller performance improvement when MAI cancellation is used. At high user capacities (above 8 users typically) the performance gets worse compared with simple one stage detection schemes. The reason for this is that when the uplink channel is saturated with other user noise, the initial estimates of the first iteration become very unstable. This causes the interference estimates of all other interfering users to give a "dirty" signal that can actually degrade the bit-error rate performance.

The LROCC coding scheme relies primarily on independent errors to yield good results and it is therefore important to provide initial good estimates on the first iteration of interference estimation in the receiver and to use strong interleaving in heavily faded channels. The use of OFDM in frequency selective channels helps to further combat the fading effects and effectively provides independent fading on each sub-channel. Combing this modulation strategy with LROCC coding provides an ultimately robust transmission design for such channels.

CONCLUSIONS

The proposed transmitter and receiver design for the uplink channel of a mobile communication system in heavy frequency selective Rayleigh fading can be used to combat the degradation effects through the use of LROC codes and controlled equalisation, providing a capacity gain of at least 25% at BER=10^{-3}. Of course, such gains will be degraded if the channel state information contains errors.

The use of zero forcing equalisers would not lead to a good performance without MAI cancellation, but through the use of MAI cancellation it is possible to combine zero-forcing algorithms with MRC on the final detection stage to give good results.

ACKNOWLEDGEMENTS

The authors are pleased to acknowledge the support of EPSRC for their support of this work through contract No. GR/K27902 and to NORTEL for their provision of CASE studentship support.

REFERENCES

[1] Y. Yukitoshi and M. Nakagawa, "A Multiuser Interference Cancellation Technique Utilizing Convolutional Codes and Orthogonal Multicarrier Modulation for Wireless Indoor Communications", *IEEE Journal on Selected Areas in Communications*, Vol. 14, No.8, October 1996

[2] K. Fazel, S. Kaiser and M. Schnell, "A Flexible and High Performance Cellular Mobile Communications System Based on Orthogonal Multi-Carrier SSMA", *Wireless Personal Communications*, Kluwer Academic Publishers, pp.121-144, 1995

[3] T. Mueller *et al,* "Comparison of different Detection Algorithms for OFDM-CDMA in Broadband Rayleigh Fading*", IEEE Conference Publication*, pp. 835-838, 1995.

[4] S. Kaiser, "OFDM-CDMA versus DS-CDMA: Performance evaluation for fading channels", *Proc. IEEE International Conference on Communications* (ICC'95), pp.1722-1726, June 1995.

[5] Maxey, J J and Ormondroyd, R F, "Optimisation of orthogonal low-rate convolutional codes in a DS-CDMA system", *IEEE/URSI Conference Proceedings of ISSSE'95*, Vol. 95TH8047, pp. 493-496, 1995

[6] Maxey, J J and Ormondroyd, R F, "Low-Rate Orthogonal Convolutional Coded DS-CDMA using Non-Coherent Multi-Carrier Modulation over the AWGN and Rayleigh Faded Channel", *IEEE Conference Proceedings of ISSSTA'96*, Vol. 2, pp.575-579, 1996

[7] COST 207: *Digital land mobile radio communications*, Final Report, Commission of the European Communities, Luxembourg 1989

AN APPROACH FOR A MULTI-CARRIER SPREAD SPECTRUM SYSTEM WITH RAKE RECEIVER†

Achim Nahler, Gerhard P. Fettweis

Dresden University of Technology, Mobile Communications Systems
D-01062 Dresden, Germany, Tel.: +49 351-463 5521, Fax.: +49 351-463 7255
e-mail: nahler@ifn.et.tu-dresden.de

ABSTRACT

Multi carrier spread spectrum is a new kind of modulation technique and can be viewed as the dual technique to direct sequence [1]. Hence many investigations have been focussing on detection in the frequency domain [2], [3]. However, we analyze the multi carrier spread spectrum (MC-SS) signal in the time domain. In out paper we compare time and frequency domain properties of both techniques, DS-SS and MC-SS and prove that the auto correlation function (ACF) of a MC code signal is independent from the code sequence in contrast to a DS code signal. Finally, we will show that the behavior of ACF allows the use of a time domain RAKE receiver for MC like it is used many DS systems.

I. INTRODUCTION

It is well known that DS-SS has a sinc-shaped spectrum while MC-SS has a rectangular-shaped spectrum. A system with rectangular shaped spectrum has two advantages over a sinc-shaped spectrum: 1. It is more resistant against fading and interference effects. 2. Out of band power decreases faster than the

†This work is supported by the German National Science Foundation (Deutsche Forschungsgemeinschaft contract Fe 423/1)

K. Fazel and G.P. Fettweis (eds.), Multi-Carrier Spread-Spectrum, 97-104.

out of band power of a sinc-shaped spectrum. Thus, filtering is necessary for DS-SS system. Since filtering has to be realized at the chip level it leads to hard real-time demands with a non-negligable cost factor.
RAKE receivers are often used for combatting multipath frequency-selective fading effects which are typical for mobile communication channels. This path diversity technique provides large performance enhancement. A further advantage of the RAKE is its low complexity and easy implementation. Many DS-SS systems (e. g. IS-95) use a RAKE because of their spreading sequences' autocorrelation properties.
This paper is organized as follows. Section II describes DS-SS and MC-SS in time and frequency domain and compare them. In Section III we examine the ACF of MC-SS in erspect to DS-SS code signals in the time domain and in Section IV we finally show some simulation results of a RAKE-based MC-SS system.

II. DIRECT SEQUENCE VERSUS MULTI CARRIER

In [4] and [2] the idea of multi carrier spread spectrum modulation was introduced. A data symbol $d(t)$ of duration T_S is modulated on N subcarriers which are orthogonal to each other with a subcarrier spacing $\frac{1}{T_S}$. Each of this subcarrier signals is multiplied by a complex valued scalar c_n, where all of this scalars have the same magnitude, e. g. $|c_n| = \frac{1}{\sqrt{N}}$ to normalize the power. Finally, all subcarrier signals are added. Fig. 1 also shows the MC-SS principle. In the time domain, the formulation of the transmitted signal $mc(t)$ is as follows

$$mc(t) = \sum_{n=0}^{N-1} d(t) c_n e^{j2\pi n \frac{t}{T_S}} = d(t) \cdot c_{MC}(t) \tag{1}$$

where $c_{MC}(t)$ is periodical with T_S:

$$c_{MC}(t) = \sum_{k=-\infty}^{+\infty} c_{mc}(t - kT_S) = c_{mc}(t) \star \sum_{k=-\infty}^{+\infty} \delta(t - kT_S) \tag{2}$$

where $\star$ denotes convolution and

$$c_{mc}(t) = \begin{cases} \sum\limits_{n=0}^{N-1} c_n e^{j2\pi n \frac{t}{T_S}} & 0 \leq t \leq T_S \\ 0 & \text{elsewhere.} \end{cases} \tag{3}$$

We obtain the MC spectrum as

$$C_{MC}(f) = \sum_{n=0}^{N-1} c_n \, \delta\left(f - \frac{n}{T_S}\right). \tag{4}$$

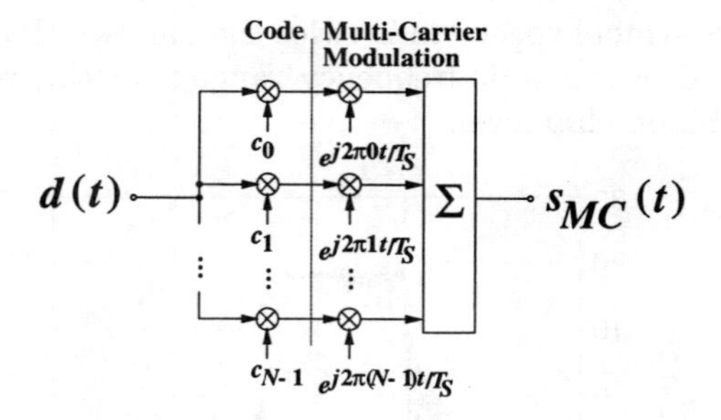

Fig. 1: Multi Carrier Spreader

It is well known that in a DS system the spread spectrum results from multiplication of data symbols $d(t)$ by a higher rated (most PN) code sequence $c_{DS}(t)$: $ds(t) = d(t) \cdot c_{DS}(t)$. We assume that code sequence $c_{DS}(t)$ is periodical with symbol duration T_S, hence

$$c_{DS}(t) = \sum_{k=-\infty}^{+\infty} c_{ds}(t - kT_S) = c_{ds}(t) \star \sum_{k=-\infty}^{+\infty} \delta(t - kT_S) \tag{5}$$

where $c_{ds}(t)$ is

$$c_{ds}(t) = \begin{cases} \sum\limits_{n=0}^{N-1} c_n \cdot \text{rect}\left(\frac{Nt}{T_S/2} - n + 1\right) & 0 \leq t \leq T_S \\ 0 & \text{elsewhere} \end{cases} \tag{6}$$

with $c_n \in \{-1; 1\}$ and the chip duration $\frac{T_S}{N}$. The DS spectrum is given by

$$C_{DS}(f) = \text{sinc}\left(\pi f \frac{T_S}{N}\right) \cdot \sum_{k=-\infty}^{+\infty} \delta\left(f - \frac{k}{T_S}\right) . \tag{7}$$

Now we can compare MC-SS with unfiltered DS-SS. We assume the same spreading factor and same symbol duration for both systems. Then both systems have the same subcarrier spacing (Eq. (4) and (7)), exactly. We can interprete DS-SS as MC-SS with infinite number of subcarriers (7)). And, the envelope of these subcarrier magnitudes is a sinc-function in contrast to MCSS, where all subcarriers have the same magnitude and only differ in phase offsets. On the other hand, MC-SS can be viewed as DS-SS with a complex valued code sequence instead of a code sequence having only the values 1 and -1. Furthermore a binary symbol modulated by MC-SS has far better spectral properties than modulated by DS-SS without filtering (see Fig. 2). The most power is concentrated in the desirable frequency range, because the MC-SS symbol spectrum is rectangular shaped. The null-to-null-bandwidth of a sinc-shaped DS-SS symbol spectrum is twice the bandwidth of MC-SS and the power is distributed over a wide frequency range. For this reasons filtering is necessary for DS-SS on chip level.

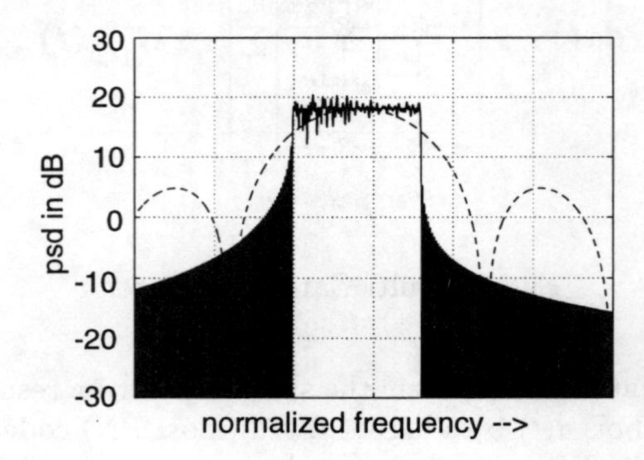

Fig. 2: Spectrum of a binary symbol modulated by MC-SS (–) and DS-SS (- -)

III. AUTO CORRELATION FUNCTION OF SPREAD SPECTRUM CODE SIGNALS

In this section we investigate the cyclic ACF (CACF) of the MC-SS and DS-SS code signal. The CACF is defined as

$$\varphi(t) = \int_{-T_S}^{T_S} x^*(\tau)x(t+\tau)\, d\tau \tag{8}$$

where x is a T_S-periodical signal. At first we examine the MC-SS code signal (see Eq. (2) and (3)). Considering the subcarriers' orthogonality we obtain for the CACF

$$\varphi_{MC}(t) = const. \cdot \sum_{n=0}^{N-1} e^{j2\pi n \frac{t}{T_S}} \ . \tag{9}$$

The real scalar *const.* depends only on the spreading factor N and the symbol duration T_S. We note that the behavior of the CACF is independent from the spreading code sequence. Fig. 3 shows the magnitude of the normalized CACF for $N = 16$ and $N = 1024$. For detection algorithms it is important that the MC-SS code signal has a narrow peak and small sidelobes in both, the frequency and the time domain.

For the DS-SS code signal (Eq. (5) and (6)) the CACF follows

$$\varphi_{DS}(t) = \sum_{n_1=0}^{N-1} \sum_{n_2=0}^{N-1} c_{n_1} c_{n_2} \cdot \Delta \left(\frac{Nt}{T_S/2} + n_1 - n_2 - 1 \right) \tag{10}$$

with

$$\Delta(t) = \begin{cases} \frac{t}{2} + 1 & \text{if } -2 \leq t \leq 0 \\ -\frac{t}{2} + 1 & \text{if } \quad 0 \leq t \leq 2 \\ 0 & \text{elsewhere.} \end{cases}$$

Eq. (10) shows that the exact behavior of a DS-SS code signal's CACF depends on the code sequence itself in contrast to MCSS, where the CACF is independent from the code signal.

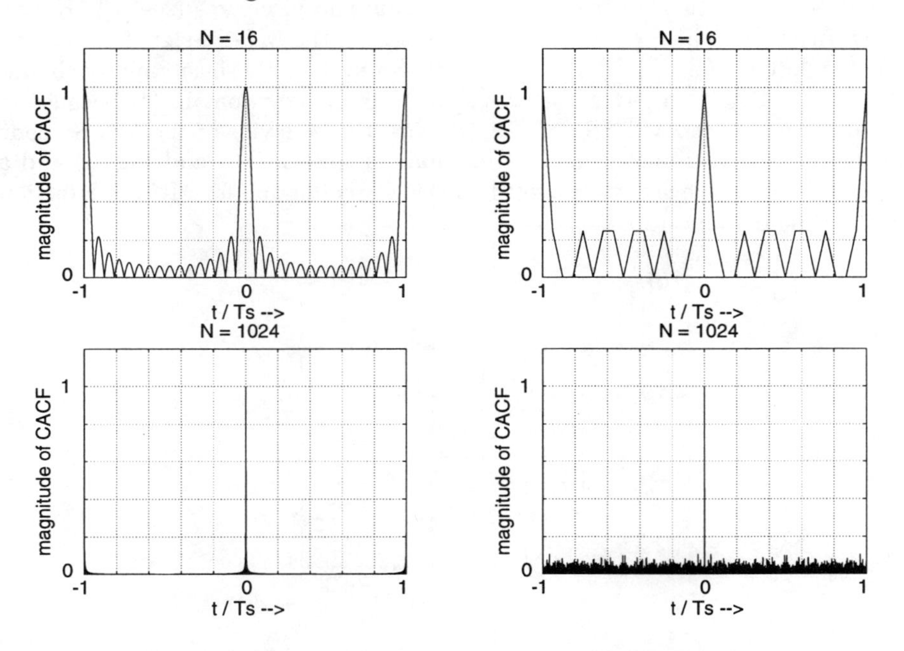

Fig. 3: CACF of MC-SS code signal   Fig. 4: CACF of DS-SS PN code signal

IV. RAKE RECEIVER AND SIMULATION RESULTS

Time-varying frequency- and/or time-selective multipath propagation is typical for mobile communication channels [5]. Essentially two parameters characterize the mobile communication channel: The Delay spread T_D excited by multipath propagation and the Doppler spread f_D excited by time-variation of channel.

These parameters define the coherence bandwidth and coherence time, respectively. The Fourier transform of channel impulse response is in magnitude and phase nearly constant inside the coherence bandwidth $f_C \approx \frac{1}{T_D}$. Similarly, coherence time $T_C = \frac{1}{f_D}$ specifies the time within which the channel impulse response is slowly changing. In [6] it was proven that the RAKE receiver is the optimum receiver for frequency-selective and time-nonselective fading channels ($f_C <$ signal bandwidth and $T_C \gg$ symbol duration). A long symbol duration and a large bandwidth due to spread spectrum techniques are characteristics of spread spectrum systems. DS-SS signal detection is realized in the time domain. The ACF properties of PN code sequences in the time domain allows the use of a RAKE receiver in DS-SS systems. One can interpret MC-SS as the dual system to DS-SS such that all DS-SS operations in the time domain are MC-SS operations in the frequency domain and vice versa (see [1]). Hence, many investigations are focussed on detection of the MC-SS-signal in the frequency domain (e. g. [2], [3]). For a MC-SS system the basic approach was therefore a RAKE receiver operating in the frequency domain [1] because of the duality between MC-SS and DS-SS. Now the behavior of the MC-SS code sequences is better known in the time domain (see Eq. (9) and Fig. 3) and a practical consequence is the use of the RAKE architecture for MC-SS detection in the time domain, too.

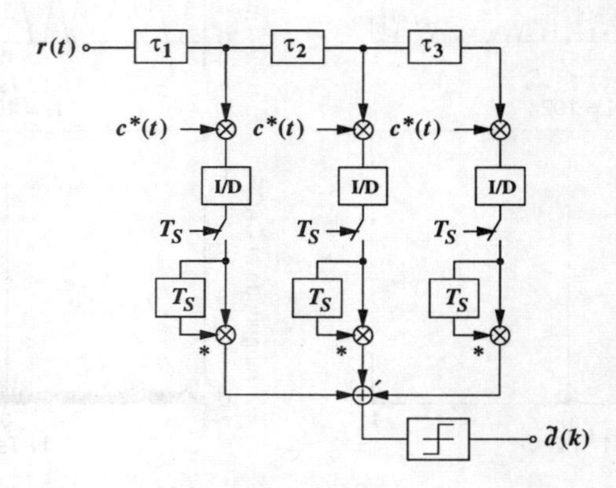

Fig. 5: RAKE receiver for noncoherent DPSK demodulation

For simulations, a spread spectrum signal bandwidth of $B = 10$ MHz and a spreading factor $N = 16$ are the basic parameters of our MC-SS system. Therefore, the symbol duration is $T_S = \frac{N+1}{B} = 1.7\ \mu$s and the subcarrier spacing $\frac{1}{T_S} = 588$ kHz. A special code sequence is used, so that code signal generated by this sequence has a reduced dynamic range of about 4.2 dB in the time domain. That allows to profit from the power efficiency of class C power amplifier. A detailed description of developing codes with reduced dynamic

Table 1: Channel characteristics (COST207 BU)

tap	tap delay $\frac{t}{T_S}$	power
1	0.00 ... 0.25	1.0000 ... 0.6538
2	1.50 ... 1.75	0.0781 ... 0.0510
3	2.50 ... 2.75	0.0143 ... 0.0093
4	3.25 ... 3.50	0.3269 ... 0.2137
5	4.50 ... 4.75	0.0390 ... 0.0255
6	5.50 ... 5.75	0.0071 ... 0.0047

range is included in [7].
The channel model used is based on a tapped delay line model as implemented in the simulation software COSSAP. We used the 6-path-model where each tap is Rayleigh-distributed according to the COST207 BU channel model (see Table 1). In addition to COSSAP we allowed for time-varying delays of each of the 6 paths. For DBPSK modulation we use a 3-finger-RAKE (see Fig. 5) which profits from 94 % to 97 % of signal power for symbol detection. We

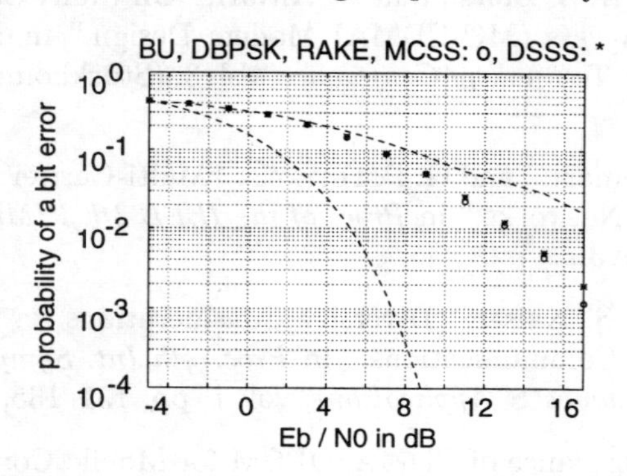

Fig. 6: Performance of MC-SS and DS-SS System

evaluated the performance of the RAKE receiver under the condition that the fading is sufficiently slow to perfectly estimate the complex valued channel impulse response. In Fig. 6 simulation results are plotted. Both dashed curves are upper and lower performance bounds, respectively. The upper bound is given by the performance of the AWGN channel. And the lower bound results from a correlation receiver, which perfectly detects only the most powerful path ("1-finger RAKE"). The MC-SS-RAKE is compared with DS-SS-RAKE using the same spreading gain and symbol duration.

V. CONCLUSIONS

We analyzed MC-SS and DS-SS in time and frequency domain. DS-SS can be viewed as MC-SS with an infinite number of subcarriers where the magnitude of subcarriers is sinc-distributed. Frequency efficiency of DS-SS without filtering is worse than MCSS. The behavior of MC-SS CACF is independent of the code signal in contrast to DS-SS. A narrow peak and small sidelobes by autocorrelation function allow the use of a time-domain-RAKE. A performance analysis shows that a RAKE-based MC-SS system is comparable to DS-SS relating to the bit error rate.

ACKNOWLEDGMENT

The authors wish to thank J. Kuehne, B. Stantchev and V. Aue for the fruitful discussions on the area of spread spectrum techniques.

REFERENCES

[1] G. P. Fettweis, A. S. Bahai, and K. Anvari, "On Multi-Carrier Code Division Multiple Access (MC-CDMA) Modem Design," in *Proc. of the 1994 IEEE Vehicular Technology Conference*, vol. 3, (Stockholm), pp. 1670–1674, 1994.

[2] N. Yee, J.-P. Linnartz, and G. P. Fettweis, "Multi-Carrier CDMA in Indoor Wireless Radio Networks," in *Proc. of the IEEE 4th PIMRC*, (Yokohama), pp. 109–113, 1993.

[3] M. Schnell and S. Kaiser, "Diversity Considerations for MC-CDMA Systems in Mobile Communications," in *Proc. 4th Int. Symposium on Spread Spectrum Techniques & Applications*, vol. 1, pp. 131–135, Sept. 1996.

[4] K. Fazel, "Performance of CDMA/OFDM for Mobile Communication Systems," in *Proc. of the IEEE 2th ICUPC*, (Ottawa), pp. 975–979, 1993.

[5] P. A. Bello, "Characterization of Randomly Time-Variant Linear Channels," *IEEE Transactions on Communications Systems*, vol. CS-11, pp. 360–393, Dec. 1963.

[6] G. Price and P. E. Green, "A Communication Techniques for Multipath Channels," *Proceedings IRE*, vol. 46, pp. 555–570, March 1958.

[7] V. Aue and G. P. Fettweis, "Higher-Level Multi-Carrier Modulation and Its Implementation," in *Proc. 4th Int. Symposium on Spread Spectrum Techniques & Applications*, vol. 1, pp. 126–130, Sept. 1996.

MULTI-CARRIER MODULATED ORTHOGONAL CODE-DIVISION MULTIPLE ACCESS (MCM-OCDMA)

D. Thomas Magill
Stanford Telecommunications, Inc.
1221 Crossman Ave.
Sunnyvale, CA 94088-3733 USA

ABSTRACT
A waveform and communication system is described which permits fixed users to operate with the advantages of OCDMA in a multipath channel.

INTRODUCTION
CDMA has become a very popular method of wireless multiple access. The IS-95 CDMA standard is a well-known approach for mobile cellular applications. IS-95 is being extended for future application to PCS and such things as wireless local loop. While IS-95 uses OCDMA for the forward link it uses asynchronous CDMA on the reverse link. Asynchronous CDMA cannot be orthogonal and is subject to multiple access interference (MAI) from the other users within the cell as well as from adjacent cells. This access noise limits the performance of the return link and the system. The MAI has two components –inter-cell and intra-cell access noise. Use of OCDMA permits the elimination of the former. In general, geometry and symbol rates preclude the possibility of synchronous OCDMA code channels with adjacent cells and, therefore, inter-cell MAI remains even with an OCDMA system.

For mobile cellular systems having users with omni-directional antennas approximately one half the MAI is intra-cell and a substantial gain is obtained by using OCDMA. For fixed applications in which the subscribers as well as the base station can employ directional antennas the inter-cell MAI contribution is much lower and even greater capacity gains can be obtained with OCDMA.

FACTORS INCREASING THE CROSSTALK BETWEEN "ORTHOGONAL" CHANNELS
There are four major factors that decrease the independence of "orthogonal" channels. First, bandwidth limiting distorts the binary orthogonal codes and causes crosstalk.

K. Fazel and G.P. Fettweis (eds.), Multi-Carrier Spread-Spectrum, 105-110.

Figure 1 illustrates the interference-to-signal power ratio as a function of the static offset in chips for the case of an 32-ary Rademacher-Walsh (RW) BPSK format with the rectangular RW chips randomized by a common PN code on a one PN chip per one RW chip basis. The desired signal has the time base error so that it sees the maximum interference. Note that due to filtering (4-pole, 0.1 dB-ripple Chebychev with the 3-dB point at the first spectral null) there is finite, but small, crosstalk even with no time base error. Second, time base errors will create crosstalk as illustrated in Fig. 1. It is clearly desirable to keep the time base error to a small percentage of a chip duration or less. In general, having the base station feedback to each subscriber the required time base correction can virtually eliminate fixed offset errors. However many wireless channels have delay spread due to multipath propagation and it is impossible for the base station to give the subscriber a time base correction that will eliminate crosstalk due to delay spread. Delay spread is the principal reason that OCDMA has not been considered feasible for the return link and is a limiting factor on the performance of the forward link. Third, nonlinear effects in power amplifiers, or other components may cause crosstalk due to the creation of IM crossproducts. Thus, it is important to operate the subscriber terminal power amplifier in a linear mode -- particularly if it is transmitting multiple RW channels. Fourth, carrier frequency error will also cause crosstalk but it is not usually a significant effect for high chip rate systems. However, when low chip rates are used to combat delay spread it is important to have a network frequency control system to minimize carrier frequency errors.

METHODS OF REDUCING THE IMPACT OF DELAY SPREAD ON OCMA PERFORMANCE

Equalization at the base station demodulator is not possible. Since each subscriber sees a different multipath channel, there is no single filter that can operate on the composite received signal so as to remove MAI. There are two principal methods of compensating for delay spread such that near orthogonal performance is achievable. First, the base station could inform the subscriber how to pre-distort its transmitted signal such that it appears as an orthogonal signal at the base station. This approach requires the base station accurately estimate the channel impulse response function and that the subscriber implements the inverse filter.

A more attractive alternative is to reduce the chipping rate such that delay spread is negligible in comparison to the chip duration. For certain applications such as a wireless PBX system operating in the ISM bands this situation may naturally arise due to the low delay spread in an indoor environment.

If one must operate in an outdoor environment and/or at high data rates, a very good way of reducing the chip rate is to use multi-carrier modulation. For example, if 32 parallel carriers or tones are used to transmit the data the chip rate can be reduced by a factor of 32. Such a reduction can result in negligible crosstalk in many applications and does not require any overhead for transmission of inverse filter parameters to each subscriber.

REPRESENTATIVE MCM-OCDMA SYSTEM

Consider an application to a fixed wireless network supporting 64 kb/s traffic with 32 orthogonal channels. Two of these channels are reserved for radio system overhead functions such as power, time, and frequency control and demand assignment. The remaining channels are available for traffic and telephone signaling. Randomizing the orthogonal RW functions with a common PN code creates the orthogonal BPSK codes.

As for all orthogonal access systems it is important for OCDMA to use a bandwidth efficient modulation system. QPSK modulation with DPSK detection is used on each of the 32 MCM tones. Thus, the basic data symbol rate is 1 ksymbol/s resulting in a 32 kchips/s chip rate. If were not for the delay spread it would be possible to place the MCM tones 32 kHz apart, which is the minimum bandwidth possible for orthogonality. However, since the delay spread would cause inter-MCM tone crosstalk it is necessary to time gate the chip detector and increase the separation of the MCM tones proportionally. A 95% integration interval for each chip results in an approximate 5% increase in the spacing of the MCM tones and corresponding decrease in system bandwidth efficiency. There is also a loss of about 0.25 dB in power efficiency in the AWGN channel. However, these minor losses are more than made up for by the insensitivity to MAI induced by delay spread. There is no MAI for delay spreads less than ± 0.73 microsecs. This is adequate for many fixed applications using directional antennas.

For synchronization purposes both the forward and return links contain overhead bursts. On the forward link the overhead burst is called the sounding burst. It consists of one symbol on each of 32 tones and is transmitted once per frame consisting of 17 symbols and of 16 ms duration. The forward link sounding burst is transmitted exclusively, i.e., all other RW channels are turned off at this time, on RW channel 0, which is the PN code. Since this is the only signal present in this time slot it is possible, if so desired, to use 15 dB more power for this signal which is used by all subscribers to achieve receive timing, transmit clock rate, and set network carrier frequencies. On the return link one symbol per frame is time shared between the subscribers on a master frame basis. These return link overhead bursts allow the base station to measure the time base error of the individual subscriber terminals. Fig. 2 illustrates the forward link frame structure.

The overhead factor of 17/16 increases both the chip and data symbol rates proportionally to 34 kchips/s and 1.0625 ksymbols/s. Fig. 3 illustrates the transmitted power spectral density of this representative MCM-OCDMA waveform. Note that 1.92 Mb/s of non-FEC coded data is supported within a first-null-to-first null bandwidth of 1.177 MHz. Such a signal can be fit into a 1.25 MHz channel resulting in a single cell bandwidth efficiency of 1.53 b/s/Hz. Table 1 summarizes the characteristics of this representative MCM-OCDMA waveform.

Fig. 4 illustrates the simulated performance of the waveform with a slightly larger, chip-blanking factor of 9.4% which corresponds to approximately a 2.7 μsec region of immunity to delay error/spread. This figure displays the fully loaded (all RW channels present at the same power level) access noise power (dB below the desired signal) for a representative (in this case the sixteenth) MCM tone. This power is shown as a function of timing error (in chips) and carrier frequency error normalized by the symbol rate, dF. There are two sets of curves presented. The "gull-winged" set corresponds to a non-gated chip detector. This detector suffers from access noise from other MCM tones induced by time base error. The "flat-bottomed" set corresponds to a gated chip detector. Use of chip gating clearly produces a waveform that is much more robust in the presence of delay error or delay spread and that is no more sensitive to carrier frequency error than the non-gated chip detector. (For zero delay error the access noise power is slightly lower for the non-gated detector since it does not suffer the mismatched filter loss of the gated chip detector.) For this example the access noise will negligible in comparison to the AWGN if the network delay and frequency errors can be kept to within ±1.37 μsec and ± 22 Hz, respectively.

DATA DEMODULATOR CONSIDERATIONS

Differentially coherent data detection (DPSK) was selected for its robust performance in fading channels and its simplicity. A single DPSK detector and AFC discriminator can be time-shared amongst all MCM tones. DPSK detection is also tolerant of group delay distortion caused by channel filtering or multipath propagation. Furthermore, it may be helpful for systems employing diversity. However, DPSK detection of QPSK does suffer approximately a 2.3-dB degradation in power efficiency as compared to coherent detection. Furthermore, many FEC coding perform poorly DPSK detection. Our interest in implementation simplicity and robustness were the principal reason for our choice of DPSK detection.

IMPLEMENTATION CONSIDERATIONS

Rectangular chips were selected for their simplicity of implementation and tolerance of time base errors (with chip gating). Furthermore, near optimal bandwidth efficiency can be achieved in a multipath channel without requiring group delay equalization across the band.

Due to the non-constant envelope of MCM a backoff penalty of about 9 dB may be required in the subscriber power amplifier. In those cases, in which the subscriber transmits simultaneously on multiple RW channels there is already a non-constant envelope so the backoff penalty associated with the use of MCM will be less. At the base station the power amplifier must already be backed off by about 9 dB to support multiple RW channels so the use of MCM encounters no additional backoff penalty. By contrast, with a 32-slot TDMA system the average-to- peak power backoff for a subscriber is 15 dB.

CONCLUSIONS

For fixed link applications, such as wireless local loop, multi-carrier modulation makes it possible to achieve orthogonal CDMA operation for both forward and return links of a star network. Use of OCDMA rather than conventional asynchronous CDMA is advantageous in that intra-cell MAI is virtually eliminated thereby increasing capacity. For fixed links using directional antennas the capacity gain is even greater due to the lower level of inter-cell MAI.

Table 1. Representative MCM-OCDMA System Characteristics

- 32-ary RW channels
- Supports 64 kbits/s traffic in each RW channel
- Use of uncoded QDPSK data modulation
- Use of 32-ary multi-carrier modulation
- Use of sounding burst for synchronization on both forward and return links
- Bandwidth of 1.177 kHz supports 30 channels of traffic and telephone signaling
- 2 RW channels used for radio control and DAMA
- Frame structure has 1 overhead symbol per frame of 17 symbols
- The chip rate (BPSK) is 34 K chips/s
- No MAI for delay spreads less than ±0.69 microsecs

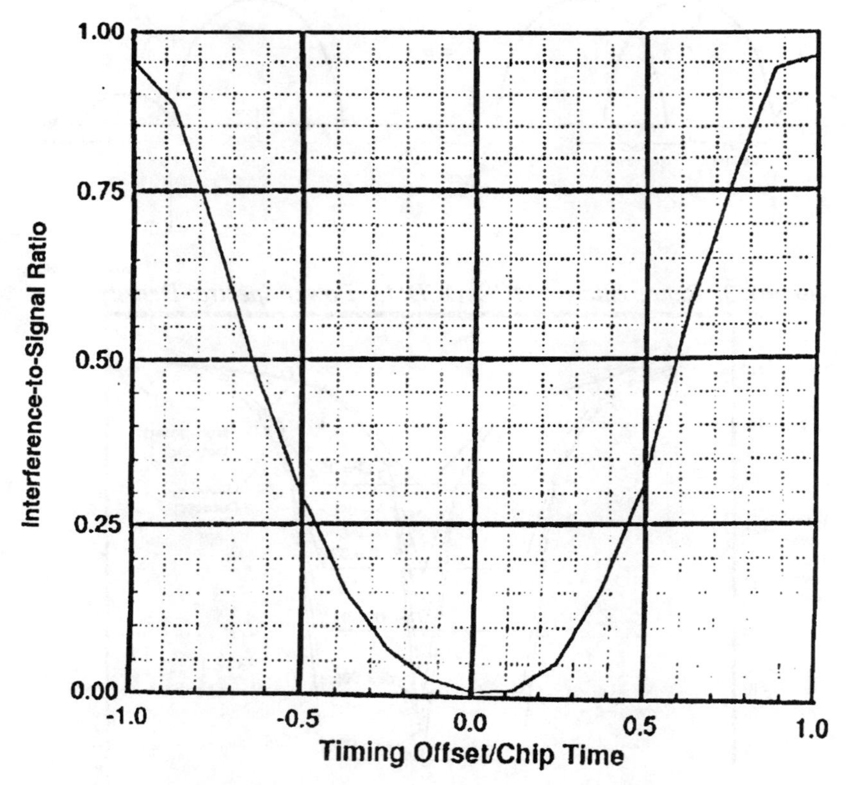

Figure 1. Access Noise-to-Signal Ratio vs Time Base Error

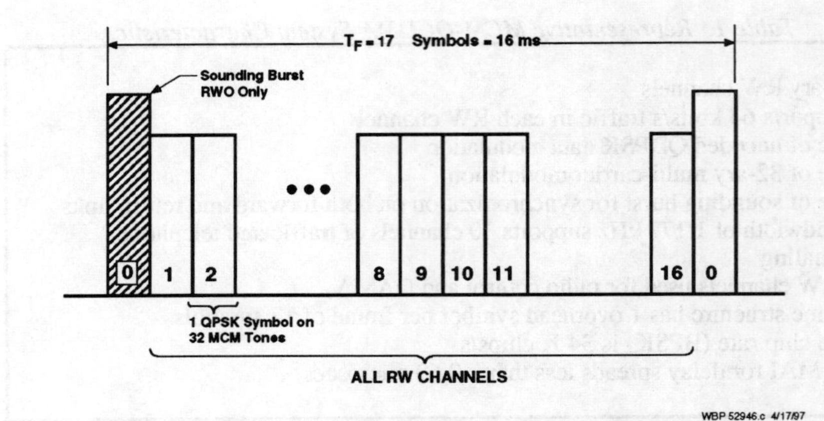

Figure 2. MC-OCDMA Frame Structure

Figure 3. Representative MCM-OCDMA Power Spectral Density

Figure 4. Total Access Noise Power vs Center Frequency and Time Errors for Non-gated Detector

INTERLEAVED FDMA – A NEW SPREAD-SPECTRUM MULTIPLE-ACCESS SCHEME

Uli Sorger[1], Isabella De Broeck[1], and Michael Schnell[2]

[1] Inst. for Network- and Signal Theory, Darmstadt Univ. of Technology, Merckstraße 25, D–64283 Darmstadt, Germany
Phone: +49 (0) 6151 16-3215, -3614 / Fax: -4313

[2] Inst. for Commun. Tech., German Aerospace Research Establ. (DLR)
P.O. Box 11 16, D-82230 Weßling, Germany
Phone: +49 (0) 8153 28-2858 / Fax: -1442

ABSTRACT

In this paper, a new spread-spectrum multiple-access scheme applicable for both the uplink and the downlink of a mobile radio communications system is proposed, which is termed "Interleaved Frequency-Division Multiple-Access". This new scheme can be viewed as a special kind of a multicarrier spread-spectrum multiple-access scheme, since each user is assigned a different set of orthogonal frequencies. The orthogonality between the different users is maintained at the receiver, even in time-dispersive channels. Thus, there is no multiple-access interference. However, the cost to avoid multiple-access interference is to allow for intersymbol interference and therefore an equalization is required at the receiver. This article describes the new spread-spectrum multiple-access scheme as well as the optimum receiver structure.

K. Fazel and G.P. Fettweis (eds.), Multi-Carrier Spread-Spectrum, 111-118.

INTRODUCTION

The three classical multiple-access schemes are "Frequency-Division Multiple-Access" (FDMA), "Time-Division Multiple-Access" (TDMA), and "Code-Division Multiple-Access" (CDMA), each having its own characteristic advantages and disadvantages. The proposed multiple-access scheme "Interleaved Frequency-Division Multiple-Access" (IFDMA) is a new spread-spectrum multiple-access (SSMA) scheme which shows several advantages in comparison with the classical multiple-access schemes. Unlike TDMA a continuous transmission is used and compared to CDMA no spreading sequence is necessary for the discrimination of the different user signals. Moreover, no multiple-access interference (MAI) arises even in a time-dispersive channel, such as the mobile radio channel. However, the cost to avoid MAI is to allow for intersymbol interference (ISI). Thus, IFDMA requires an equalization at the receiver in the case of transmitting over a time-dispersive channel. In contrast to TDMA and CDMA, each user in an IFDMA system is assigned a different set of orthogonal frequencies. Thus, IFDMA can be viewed as a special kind of a multicarrier SSMA scheme. In comparison with multicarrier CDMA IFDMA offers the additional advantage of having a constant envelope. The paper is organized as follows. In Section 2, the new multiple-access scheme is proposed and described. The orthogonality between the received signals from different users is proved in Section 3, and the optimum receiver structure is given in Section 4. Finally, some conclusions are drawn.

SYSTEM DESCRIPTION

The proposed IFDMA scheme performs a blocked transmission of Q complex-valued symbols $d_q^{(i)}$, $q = 0, ..., Q-1$, for each user i, $i = 1, ..., N_u$. The symbols $d_q^{(i)}$ are assumed to be equally likely. The symbol rate is $R_s = 1/T_s$ and T_s denotes the symbol duration. In the sequel, the transmission of a single block is considered. Using vector notations a block of Q complex-valued symbols can be described according

$$\mathbf{d}^{(i)} = \left[d_0^{(i)}, d_1^{(i)}, ..., d_{Q-1}^{(i)}\right]^T \tag{1}$$

and specifies an IFDMA-symbol of user i. In order to avoid inter-block interference each IFDMA-symbol is preceded by a guard interval of length T_Δ which has to fulfill the condition

$$T_\Delta > \tau_{max} \quad , \tag{2}$$

where $\tau_{\max}$ is the maximum delay time of the transmission channel. The duration T of an IFDMA-symbol including the guard interval is

$$T = Q \cdot T_s \quad . \tag{3}$$

An IFDMA-symbol is obtained by compressing each of the Q symbols from symbol duration T_s to "chip duration" T_c and repeating the resulting compressed block $(L + L_\Delta)$–times, where the chip duration T_c is given by

$$T_c = \frac{T_s}{L + L_\Delta} \quad . \tag{4}$$

Alternatively, the construction of an IFDMA-symbol can be described as dividing each of the Q symbols into $(L + L_\Delta)$ chips followed by an additional $Q \times (L + L_\Delta)$ block interleaving. The resulting IFDMA-symbol $\mathbf{c}^{(i)}$,

$$\mathbf{c}^{(i)} = \frac{1}{L + L_\Delta} \cdot \underbrace{\left[d_0^{(i)}, \ldots, d_{Q-1}^{(i)}; \ldots; d_0^{(i)}, \ldots, d_{Q-1}^{(i)}\right]^T}_{(L + L_\Delta)\text{–times}} , \tag{5}$$

is of dimension $L_c = (L + L_\Delta)Q$ and the components $c_l^{(i)}$ can be written as

$$c_l^{(i)} = \frac{1}{L + L_\Delta} \cdot d_{l \bmod Q}^{(i)} \quad , \quad l = 0, \ldots, L_c - 1 \quad . \tag{6}$$

The first $L_\Delta Q$ chips $c_l^{(i)}$, $l = 0, \ldots, L_\Delta Q - 1$, belong to the guard interval. Thus, the duration T_Δ of the guard interval is given by

$$T_\Delta = L_\Delta \cdot Q \cdot T_c \quad , \tag{7}$$

where L_Δ has to be chosen to meet the condition of Eq. (2). The transmission signal $\mathbf{x}^{(i)}$ is constructed out of the IFDMA-symbol $\mathbf{c}^{(i)}$ by elementwise multiplication with an user dependent phase vector $\mathbf{s}^{(i)}$ of dimension L_c having components

$$s_l^{(i)} = \exp\left\{-\jmath \cdot l \cdot \Phi^{(i)}\right\} \quad , \quad l = 0, \ldots, L_c - 1 \quad . \tag{8}$$

The user dependent phase $\Phi^{(i)}$ is chosen to be

$$\Phi^{(i)} = i \cdot \frac{2\pi}{QL} \quad . \tag{9}$$

The elementwise multiplication of the two vectors $\mathbf{c}^{(i)}$ and $\mathbf{s}^{(i)}$ assures the user discrimination by assigning to each user i a different set of orthogonal frequencies. The resulting transmission signal vector $\mathbf{x}^{(i)}$ can be written as

$$\mathbf{x}^{(i)} = \left[c_0^{(i)}, c_1^{(i)} e^{-\jmath\Phi^{(i)}}, c_2^{(i)} e^{-\jmath 2\Phi^{(i)}}, \ldots, c_{L_c-1}^{(i)} e^{-\jmath(L_c-1)\Phi^{(i)}}\right]^T \quad . \tag{10}$$

Due to the symbol compression involved in the construction of an IFDMA-symbol spreading is done by the data, i.e. by the Q compressed complex-valued

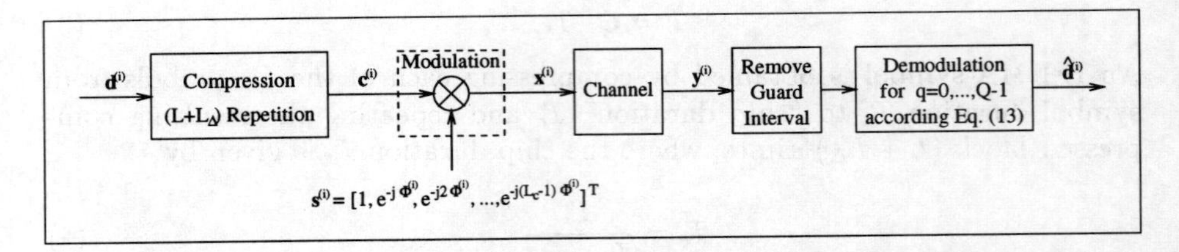

Figure 1: Block diagram of the proposed IFDMA system

symbols. For large Q the sequence of the Q compressed symbols can be assumed to be a pseudo-noise sequence and performs a task similar to that of a spreading sequence in a CDMA system. Consequently, the bandwidth B of the transmission signal $\mathbf{x}^{(i)}$ is determined approximately by the chip duration T_c

$$B = \frac{1}{T_c} \tag{11}$$

After transmission over an arbitrary channel with impulse response vector $\mathbf{h}^{(i)} = \left[h_0^{(i)}, h_1^{(i)}, ..., h_M^{(i)}\right]^T$ of dimension $M+1$, the received signal vector $\mathbf{y}^{(i)}$ of user i is given by

$$\mathbf{y}^{(i)} = \mathbf{x}^{(i)} * \mathbf{h}^{(i)} \quad , \tag{12}$$

where "$*$" specifies convolution. The received signal vector $\mathbf{y}^{(i)}$ is of dimension $L_c + M$ and the components are denoted by $y_l^{(i)}$, $l = 0, ..., L_c + M - 1$. To obtain an estimate $\hat{d}_q^{(i)}$, $q = 0, ..., Q-1$, for the transmitted symbol $d_q^{(i)}$ of user i for the proposed IFDMA system the following demodulation rule is applied

$$\hat{d}_q^{(i)} = \sum_{l=L_\Delta}^{L+L_\Delta-1} y_{lQ+q}^{(i)} \cdot e^{\jmath(lQ+q)\Phi^{(i)}} \tag{13}$$

The proposed IFDMA system is summarized in Figure 1, where symbol construction, modulation, transmission, and demodulation are shown.

PROOF OF ORTHOGONALITY

Considering a noisefree transmission over a non-distorting channel with one-dimensional impulse response vector $\mathbf{h}^{(i)} = [1]$ the received signal vector $\mathbf{y}^{(i)}$ is simply the transmission signal vector $\mathbf{x}^{(i)}$ and Eq. (13) results in

$$\hat{d}_q^{(i)} = \sum_{l=L_\Delta}^{L+L_\Delta-1} x_{lQ+q}^{(i)} \cdot e^{\jmath(lQ+q)\Phi^{(i)}} = \sum_{l=L_\Delta}^{L+L_\Delta-1} c_{lQ+q}^{(i)} = \frac{L}{L+L_\Delta} \cdot d_q^{(i)} \quad . \tag{14}$$

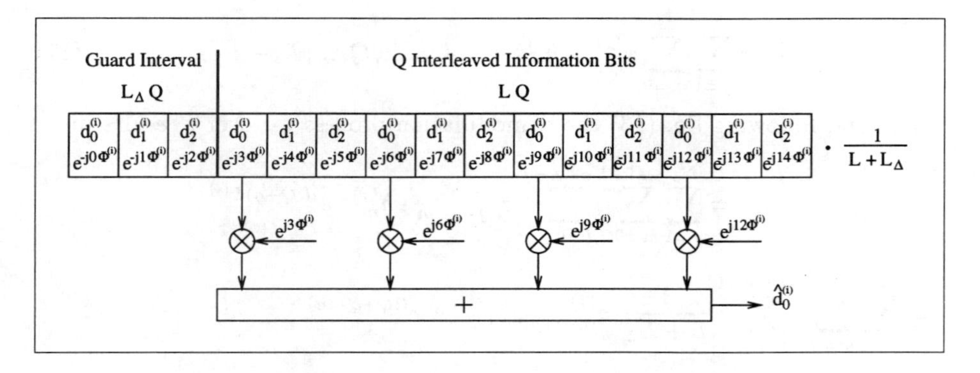

Figure 2: Illustration of the demodulation process, exemplarily for symbol $d_0^{(i)}$; $L_\Delta = 1$, $L = 4$, $Q = 3$

Thus, neglecting the scaling factor due to the guard interval the estimate $\hat{d}_q^{(i)}$ equals the transmitted symbol $d_q^{(i)}$. The demodulation process for this transmission scheme is illustrated in Figure 2. For this example the system parameters are chosen to be $L_\Delta = 1$, $L = 4$, $Q = 3$, and the demodulation for symbol $d_0^{(i)}$ is shown. Moreover, the transmitted signals $\mathbf{y}^{(i)}$ of all users i, $i = 1, ..., N_u$, are orthogonal as long as $N_u \leq L$. The demodulation of the received signal vector $\mathbf{y}^{(j)} = \mathbf{x}^{(j)}$ of an interferer j, $j \neq i$, results according Eq. (14) in

$$\sum_{l=L_\Delta}^{L+L_\Delta-1} x_{lQ+q}^{(j)} \cdot e^{\jmath(lQ+q)\Phi^{(i)}} = \sum_{l=L_\Delta}^{L+L_\Delta-1} c_{lQ+q}^{(j)} \cdot e^{\jmath(lQ+q)(\Phi^{(i)}-\Phi^{(j)})}$$

$$= \frac{d_q^{(j)}}{L+L_\Delta} \sum_{l=L_\Delta}^{L+L_\Delta} e^{\jmath(lQ+q)(i-j)\frac{2\pi}{QL}} = 0 \tag{15}$$

and it becomes appearent that the proposed IFDMA system uses orthogonal transmission signals $\mathbf{x}^{(i)}$, $i = 1, ..., N_u$. Thus, no MAI occurs in channels without time-dispersion.

In the following it is shown, that even in a time-dispersive channel with impulse response vector $\mathbf{h}^{(i)}$ of dimension $M+1$ the orthogonality of the received signals $\mathbf{y}^{(i)}$, $i = 1, ..., N_u$, is maintained as long as $N_u \leq L$. The overall received signal vector $\mathbf{y}$ for the case of N_u simultaneously transmitting users is

$$\mathbf{y} = \sum_{j=1}^{N_u} \mathbf{y}^{(j)} . \tag{16}$$

The components y_l of the overall received signal vector $\mathbf{y}$ which are relevant for the demodulation process, i.e. the components after the guard interval, can be determined according to Eq. (12)

$$y_l = \sum_{j=1}^{N_u} \sum_{m=0}^{M} x_{l-m}^{(j)} \cdot h_m^{(j)} \quad , \quad l = L_\Delta Q, ..., L_c - 1 \quad . \tag{17}$$

Therefore, following Eq. (13) the demodulation process for user i results in

$$\hat{d}_q^{(i)} = \sum_{j=1}^{N_u} \sum_{m=0}^{M} \sum_{l=L_\Delta}^{L+L_\Delta-1} x_{lQ+q-m}^{(j)} \cdot h_m^{(j)} \cdot e^{\jmath(lQ+q)\Phi^{(i)}}$$

$$= \sum_{j=1}^{N_u} \sum_{m=0}^{M} \sum_{l=L_\Delta}^{L+L_\Delta-1} \frac{1}{L+L_\Delta} d_{(q-m) \bmod Q}^{(j)} \cdot e^{-\jmath(lQ+q-m)\Phi^{(j)}} \cdot h_m^{(j)} \cdot e^{\jmath(lQ+q)\Phi^{(i)}}$$

$$= \frac{1}{L+L_\Delta} \sum_{j=1}^{N_u} \sum_{m=0}^{M} d_{(q-m) \bmod Q}^{(j)} \cdot h_m^{(j)} \cdot e^{\jmath m\Phi^{(j)}} \sum_{l=L_\Delta}^{L+L_\Delta-1} e^{\jmath(lQ+q)(\Phi^{(i)}-\Phi^{(j)})}$$

$$= \begin{cases} \dfrac{L}{L+L_\Delta} \displaystyle\sum_{m=0}^{M} d_{(q-m) \bmod Q}^{(i)} \cdot h_m^{(i)} \cdot e^{\jmath m\Phi^{(i)}} & \text{if} \quad j = i, \\ 0 & \text{otherwise.} \end{cases} \tag{18}$$

Consequently, the vector of estimates $\hat{\mathbf{d}}^{(i)}$ is a cyclic convolution "$\circledast$"

$$\hat{\mathbf{d}}^{(i)} = \frac{L}{L+L_\Delta} \mathbf{d}^{(i)} \circledast \tilde{\mathbf{h}}^{(i)} \tag{19}$$

between the vector $\mathbf{d}^{(i)}$ of complex-valued symbols and the modified channel impulse response vector $\tilde{\mathbf{h}}^{(i)}$ with phase-shifted components according to

$$\tilde{\mathbf{h}}^{(i)} = \left[h_0^{(i)}, h_1^{(i)} e^{\jmath\Phi^{(i)}}, h_2^{(i)} e^{\jmath 2\Phi^{(i)}}, ..., h_M^{(i)} e^{\jmath M\Phi^{(i)}}\right]^T \quad . \tag{20}$$

Eq. (18) shows that the orthogonality of the received signals $\mathbf{y}^{(i)}$, $i = 1, ..., N_u$, is maintained even in a time-dispersive channel. Thus, there is no MAI within a single cell of a cellular mobile radio communications system and IFDMA is an orthogonal transmission scheme like TDMA or FDMA. Note, this statement is valid for time-variant channels, but only if the channel impulse responses can be assumed to be time-invariant within one IFDMA-symbol duration T. If this assumption is violated the orthogonality of the received signals is no longer maintained resulting in MAI. The degradation due to MAI depends on the time-variance of the transmission channel.
Furthermore, the proposed IFDMA scheme is designed even to avoid worst case interference from neighbouring cells resulting in an improved frequency reuse compared to TDMA or FDMA. This is achieved by choosing the block lengths Q_μ and Q_ν for transmission in neighbouring cells μ and ν to be mutually prime

$$\gcd(Q_\mu, Q_\nu) = 1 \quad . \tag{21}$$

As a consequence, the inter-cell interference is "averaged" and can be assumed to be gaussian but not necessarily white. Note, the choice of the block length changes neither the symbol rate R_s nor the transmission bandwidth B. Only the duration T of the IFDMA-symbol is influenced according to Eq. (3).

OPTIMUM RECEIVER STRUCTURE

Considering Eq. (18), it can be seen that ISI is present at the receiver of an IFDMA system and therefore an equalization is necessary. If the impulse response is known at the receiver, the optimum equalizer can be applied which is the "Maximum-Likelihood Sequence Estimator" (MLSE) [For72].
Unfortunately, the standard Viterbi algorithm [Vit67] can not be used to implement the optimum equalizer at the receiver of an IFDMA system. This is due to the cyclic convolution between the vector $\mathbf{d}^{(i)}$ of complex-valued symbols and the modified channel impulse response vector $\tilde{\mathbf{h}}^{(i)}$ according to Eq. (18). This cyclic convolution requires a cyclic trellis representation. A cyclic trellis is a trellis without having a defined initial state. In the considered case, this is equivalent to the fact that there are no symbols which are not affected by other symbols. In order to use the Viterbi algorithm for equalization there are two alternatives. The optimum approach using the Viterbi algorithm is to set up a trellis taking into account the unknown initial state. However, the resulting trellis is of considerably higher state complexity. The second approach using the Viterbi algorithm is suboptimum but less complex. The cyclic trellis is used and multiple consecutive runs of Viterbi equalization are performed. This approach is known for the equivalent situation of "tail biting" convolutional codes [MaW86] and provides good performance.
Corresponding considerations apply if suboptimum equalizer structures are used, such as the "Decision Feedback Equalizer" (DFE) [Pro89, Qur85].

SUMMARY AND CONCLUSIONS

In this paper, we have described IFDMA, a new SSMA scheme, which is applicable for both the uplink and the downlink of a mobile radio communications system. IFDMA can be viewed as a special kind of a multicarrier SSMA scheme, since each user is assigned a different set of orthogonal frequencies.
Even in a time-dispersive channel the orthogonality between the different users is maintained at the receiver. Thus, IFDMA is an orthogonal transmission scheme and there is no MAI. In comparison to FDMA, TDMA, and multicarrier CDMA IFDMA offers additional advantages, such as a constant envelope and continuous transmission. However, IFDMA suffers from ISI and an equalization is necessary.

References

[For72] Forney, G.D.: *"Maximum-Likelihood Sequence Estimation of Digital Sequences in the Presence of Intersymbol Interference"*, IEEE Transactions on Information Theory, Vol. 18 (1972) No. 3, pp. 363-378

[Vit67] Viterbi, A.J.: *"Error Bounds for Convolutional Codes and an Asymptotically Optimum Decoding Algorithm"*, IEEE Transactions on Information Theory, Vol. 13 (1967) No. 2, pp. 260-269

[MaW86] Ma, H.H.; Wolf, J.K.: *"On Tail Biting Convolutional Codes"*, IEEE Transactions on Communications, Vol. 34 (1986) No. 2, pp. 104-111

[Pro89] Proakis, J.G.: *"Digital Communications"*, Second Edition, New York: McGraw-Hill Inc. 1989

[Qur85] Qureshi, S.U.H.: *"Adaptive Equalization"*, Proceedings of the IEEE, Vol. 73 (1985) No. 9, pp. 1349-1387

ASPECTS ON WIDEBAND MULTI-CARRIER COMMUNICATION

Jan E.M. Nilsson
Sweden's Defence Research Establishment (FOA 72)
BOX 1165, S-581 11 Linköping, Sweden,
E-mail: jann@lin.foa.se

ABSTRACT — Wideband multi-carrier, or orthogonal frequency division multiplex (OFDM), for HF communication is explored. In particular, the behavior of a system without guard-time is investigated. To combat channel distortion and noise, DBPSK and powerful coding are used. Finally, the deficiency of DPSK in a coded system, on noisy channels, is considered.

INTRODUCTION

We investigate a 125 kHz wideband OFDM system with differential binary PSK (DBPSK) modulation on each carrier. The number of carriers, and the OFDM symbol time are: 1024 and 8.2ms, respectively. The proposed system is intended to use only DBPSK and powerful coding to combat distortion and noise.

The HF channel is a troublesome channel with distortion due to multipath propagation, interfering signals, impulsive and atmospheric noise. Here we consider problems due to multipath propagation. How to handle (suppress) interfering signals has been treated earlier [4]. Based on wideband measurements we define a set of channels, representing different channel conditions. The more difficult channels have many paths with high Doppler shifts and long delays.

Firstly, we investigate the cost of removing the guard-time. On HF channels, a long guard-time is required (around 4 ms has been proposed in the literature). By instead using the guard-time redundancy to improve the coding, we show that the guard-time can be removed on an HF channel. Clearly, whether we loose or gain by removing the guard-time is highly dependent on the system and the channel conditions. Secondly, the deficiency of DPSK is investigated. In our investigations, correct synchronization, is assumed.

K. Fazel and G.P. Fettweis (eds.), Multi-Carrier Spread-Spectrum, 119-126.

PRELIMINARIES

To simplify the description a discrete baseband vector model is used. We let $\mathbf{x} = x_0, x_1, ..., x_{N-1}$, $x_k \in R^2$, denote the transmitted sequence in the time-domain such that $\mathbf{X} = X_0, X_1, ..., X_{N-1}$, $X_n \in R^2$, denotes the corresponding sequence in the frequency-domain. In the following, capital letters denote "frequency" and small letters "time". In OFDM systems, encoding or spreading, and modulation are performed in the frequency-domain, see [1, 4]. The sequence $\{X_1, X_2, ..., X_{N-1}\}$ is called an OFDM symbol and X_n a phasor. An HF channel is a slow fading channel over time, and to avoid accurate tracking of the phase variations DBPSK is selected as modulation scheme [1, p.278]. That is, assume +1.0 was transmitted on phasor n in the last OFDM symbol, then $X_n = +1.0$ if the nth data bit is zero (no change), and $X_n = -1.0$ if the nth data bit is one.

The data sequence we transmit is a codeword in a low rate error-correcting code obtained from m-sequences [2]. Then, a binary simplex code with parameters (n=1023,k=10,d=512) can be constructed, where n is the length, k is the number of information bits and d is the minimum Hamming distance [2, p.31]. Since we want a code, simple to decode, of length 1024 we have selected the orthogonal (1024,10,512) code obtained from this simplex code if one extra redundancy bit is added. In fact, this code consists of half of the codewords in the first-order Reed-Muller (1024,11,512) code [2, p.31]. To minimize the BER after decoding we use the soft-decision optimum "symbol-by-symbol maximum a-posteriori" MAP algorithm [1, p.254]. It is a very complex algorithm, but feasible here because of the very low code rate.
When we as a reference consider the bit error rate (BER) of an uncoded system we simply compare the transmitted and detected data sequence. As all transmitted bits then are assumed to be information bits we get BER as a function of signal to noise ratio (SNR) per information bit.

THE CHANNEL MODEL

The HF channel model can be divided into two parts; distortion due to multipath propagation, and noise. Wideband HF channel modeling is a difficult task, see [6]. If we want to investigate PSK and coherent detection, an accurate description of phase dependence upon frequency, is desirable. Here, however, since we employ DBPSK and non-coherent detection, a simplified "Waterson-type" channel model is used [6]. The time variant impulse response model of a multipath channel is defined as

$$h(t) = \sum_{i=0}^{M-1} \delta(t - \tau_i) a_i e^{j(2\pi f_{Di}(t)t + \phi_i)},$$

where; M=the total number of paths; a_i=relative amplitude of path i; $f_{Di}(t)$ =the Doppler frequency shift at time t of path i; ϕ_i=relative starting phase of path i; τ_i=relative path delays of path i. The energy of the strongest received path, divided by the energy of all received paths, is denoted η. On HF channels the conditions vary slowly and our approach is to select many sets of parameters which represent different conditions. Here we are only able to provide some of the results, based on the 4 channels defined in the Appendix. The wide-band background (or atmospheric) noise is modeled as additive white Gaussian noise (AWGN). A sample of the interfering signals and noise, is denoted n_k. Notice, we do not consider impulse noise and interfering signals.

To simplify the further analysis, and relate the channel model to the OFDM system, let us denote the maximum delay spread in terms of samples by ν. Then we get the discrete impulse response for sample $l = 0, 1, ..., \nu - 1$, and the discrete frequency response for phasor $n = 0, 1, ..., N - 1$, as

$$h_l = a_l e^{j(2\pi f_{D_l}(t)l+\phi_l)}; \quad H_n = \frac{1}{\sqrt{N}} \sum_{l=0}^{\nu-1} h_l e^{2\pi jln/N},$$

respectively. Notice that the multipath channels defined in the Appendix give $\nu \gg M$ and that most h_l are zero (i.e., we have a high time resolution). Now the received time-domain signal can be represented as (n_k denotes the noise)

$$y_k = \sum_{l=0}^{\nu-1} h_l x_{k-l} + n_k, \quad k = 0, 1, ..., N - 1. \tag{1}$$

SELECTION OF MODEM PARAMETERS

Let us denote by Δ_f the frequency one tone (carrier) occupies and by Δ_t the time one block ("phasor") occupies. Then $\Delta_f \Delta_t = 1$. The system (OFDM symbol) bandwidth BW is given by $N\Delta_f$. To limit inter-symbol interference (ISI) Δ_t should not be chosen to short. One reason to use OFDM is to avoid a complex time-domain equalizer. On the other hand, to mitigate distortion due to fading, in one symbol period, Δ_t should not be chosen to long. For the HF channel a Δ_t around 5-10ms seems reasonable, see [3].

A large bandwidth gives good performance, but requires a complex receiver. Therefore, a too large bandwidth is not sensible. A frequency diversity of about 6-8 is often considered enough, see [1, p.776]. To get diversity eight, we need eight times as large system bandwidth as coherence bandwidth. The coherence bandwidth of the HF channel can vary considerably. In extreme cases it can be as great as 1MHz, but most of the time it is between 100Hz-20kHz. Based on those reflections we have selected the following parameters; $\Delta_t \approx 8.2ms$ such that $\Delta_f \approx 122Hz$; $N = 1024$ giving $BW = 125kHz$.

THE RECEIVER

Assume that the delay spread of the discrete channel impulse response is at most ν and that the impulse response is time-invariant (no Doppler Spread). Then, after the FFT at the receiver, we have

$$Y_n = H_n X_n + J_n + N_n, \quad n = 0, 1, ..., N-1, \tag{2}$$

where J_n is the inter-symbol interferences (ISI) distortion over frequency and N_n the noise. Let us assume we synchronize correctly, that is, on the strongest received path and that this path is the first. Then $| h_0 |>| h_l |$, $l = 1, 2, .., \nu-1$. Furthermore, let us assume that all h_l are independent complex variables with uniform phase distributions. Next, we define $d_m = \tilde{x}_{N-m} - x_{N-m}$, where $\tilde{x}_{N-m}$ is the $(N-m)$th symbol of the previous block. Let $z = e^{2\pi jn/N}$. Then

$$J_n = \frac{1}{\sqrt{N}} \sum_{m=1}^{\nu-1} (d_m z^{-m} \sum_{k=m}^{\nu-1} h_k z^k), \quad n = 0, 1, ..., N-1.$$

Cyclic extension implies that $d_m = 0$ $(m = 1, 2, .., \nu-1)$ and we get $J_n = 0$. Here, we leave out the details, but for the codes we use and the HF channel we get $\langle | J_n | \rangle < \langle | H_n | \rangle$, where $\langle ... \rangle$ denotes the average over many OFDM symbols. The channel introduces noise proportional to the bandwidth. On the other hand, ISI introduces noise proportional to the signal power, regardless of bandwidth. In our wideband system, where the bandwidth is much larger than the number of transmitted information bits per second, channel noise dominates. Furthermore, the ISI distortion can vary considerably between different phasors (ISI distortion diversity over frequency).

Clearly, in practice, the impulse-response is not time invariant. Due to frequency (or Doppler) shift of different received paths, the received signal is disturbed by both Doppler shift and Doppler spread. To mitigate these disturbances a simple Doppler-shift corrector is implemented. However, the Doppler Spread is not removed.

IS A GUARD-TIME REQUIRED?

As guard-time, cyclic extension is used. We assume correct synchronization, i.e., on the strongest received path. Notice that the first received path is not necessarily the strongest. Hence, a guard-time is required both before and after the transmitted signal. When we compare the OFDM system with and without guard-time, enough guard-time is appended such that all ISI is removed (J_n is zero in (2)). Notice that when the channel introduces Doppler Spread, it is not cancelled by the guard-time.

To make the comparison simple, and independent of how the extra guard-time redundancy is used, we simply say the guard-time cost is 1.7dB in terms of SNR for channel 1 to 3. The data rate is reduced with a factor 2/3. That is,

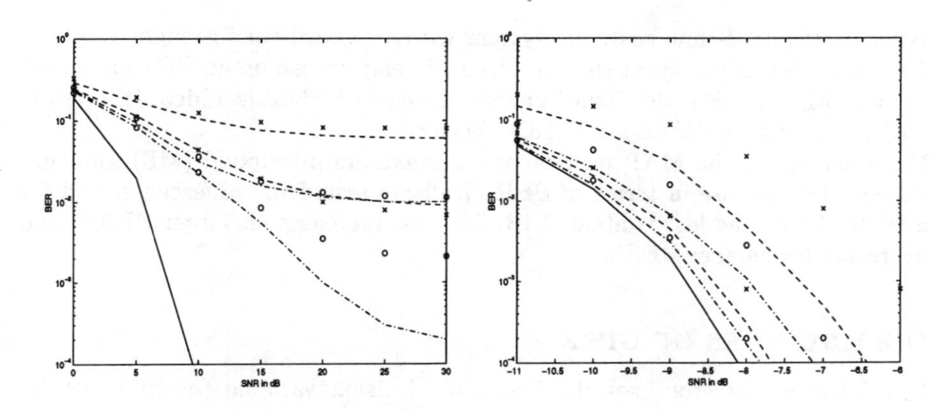

Figure 1: Uncoded and coded performance in the left and right figure, respectively: the AWGN channel: solid line; channel 1: dash-dotted line; channel 2: circles; channel 3: dashed line; channel 4: x-marks; for further explanations see the text.

we assume a guard-time of 4.1ms is enough, see [3]. However, for channel 4 a guard-time of 10.2ms would be required, i.e., a guard-time cost of 3.5dB. In the figures the system with guard-time is always better. That is, the guard-time costs are not added when the performance differences are shown. In the following, channel 0 is called the AWGN channel.

To the left in Fig.1 the uncoded BER is shown for different channels. Of two lines of the same line type, the lower (better BER) shows the result with guard-band, and the upper (worse BER) shows the result without guard-band, respectively. The system without guard-time always suffers from an irreducible bit error rate (BER) for high SNR. However, this BER is below 0.1 for all our investigated channels. With guard-time, and without Doppler-spread, the BER tends to zero for high SNR's. In Fig.1, the system with guard-time also has an irreducible BER since Doppler-spread is present. The difference with and without guard-time, at a BER of 0.1, varies but is below 5 dB, except for channel 4. However, the differences increase for higher SNR's. System degradation due to multipath propagation depends on the delay spread, number of paths, the Doppler spread, and η, where η is the energy of the strongest received path divided by the energy of all received paths.
To the right in Fig.1 the BER of the coded systems is shown for the same channels as before. The BER is plotted as a function of SNR per transmitted bit. As we can see, without guard-time the performance is close to the performance with guard-time, a loss of at most 1.5dB for channel 1 to 3. Hence, if we add

the 1.7 dB guard-time cost, the system without guard-time is slightly better. Channel 4 is really a worst case, in terms of delay spread on an HF channel [5]. On the other hand, if the "true" guard-time cost of 3.5dB is added, the system without guard-time is about "1.5dB" better.
If we instead of the MAP decoder use a maximum-likelihood (ML) sequence decoder [1], we lose in terms of BER. We have tested the differences, and for a BER of 0.01 the loss is about 1dB. The loss increases for larger BER's, and decreases for smaller BER's.

DEFICIENCIES OF DPSK

Our DBPSK detector uses the "normal" decision variable (metric), see [1, p.268]. On AWGN channels, we get

$$\alpha_n = Re(Y_n \tilde{Y}_n^*) = Re(\mathcal{E}_s + \sqrt{\mathcal{E}_s}(N_n + \tilde{N}_n^*) + N_n \tilde{N}_n^*), \tag{3}$$

where $\mathcal{E}_s$ is the symbol energy, and $\tilde{Y}_n = \tilde{X}_n + \tilde{N}_n$ the nth phasor of the previous received OFDM symbol. This is also the metric used on our multipath channel. Notice that the metric includes a non-Gaussian "noise-product" term $N_k \tilde{N}_k^*$. Thus, the decision variable for the coded case would be obtained by adding together α_n's including non-Gaussian noise. In [1] it is assumed the influence of the "noise-product" term can be neglected for SNRs of practical interest. Let us consider spread-spectrum transmission over a channel with a noise variance larger, or much larger, than the signal energy. Let $z = n * \tilde{n}$, where $n, \tilde{n}$ are two independent Gaussian noise components with zero mean and variance σ^2. Then the probability density function for z is $f_Z(z) = K_0(z/\sigma^2)$, where $K_0()$ is the zero order, modified Bessel function.

To the left in Fig.2, $f_N(n)$ and $f_Z(z)$ are shown when the noise is fairly strong; $\sigma^2 = 8$. For $f_Z(z)$ the value of the tail decreases very slowly. Hence, very large metric values, occur fairly often. Thus, a large DPSK metric value is unreliable compared to a large BPSK metric value. The problem is the noisy reference.

To the right in Fig.2, the difference (about 10dB) between normal DPSK detection (3) and the DPSK reference, for a repetition code of length 1024 is shown. The DPSK reference is the result if uncoded DPSK and repetition coded DPSK give exactly the same BER in terms of SNR per information bit. This would be true for PSK. In sequence differential detection the following metric is used. Let $Y_n = A_n + jB_n$, $\tilde{Y}_n = \tilde{A}_n + j\tilde{B}_n$. Then, for the codewords c_1, c_2 we calculate,

$$M(c_m) = \frac{1}{4}((\sum_{k=0}^{N-1} A_n - \tilde{A}_n(c_{m_k}))^2 - (\sum_{k=0}^{N-1} B_n - \tilde{B}_n(c_{m_k}))^2), m = 1, 2;$$

where, the codeword bit c_{m_n} controls if, $\tilde{A}_n(c_{m_n})$ is equal to $\tilde{A}_n$ or $-\tilde{A}_n$, and if

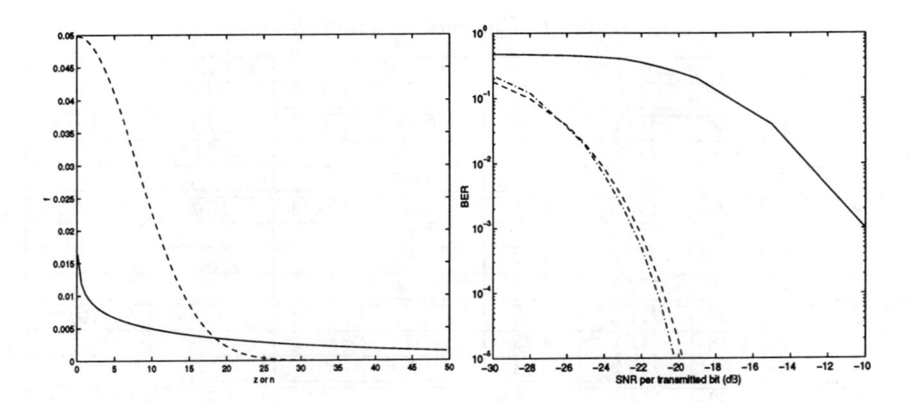

Figure 2: The left figure: $f_Z(z)$: solid line; $f_N(n)$: dashed line. The right figure shows the performance on an AWGN channel: normal DPSK detection: solid line; DPSK reference: dashed line; sequence differential detection: dash-dotted line.

$\tilde{B}_n(c_{m_n})$ is equal to $\tilde{B}_n$ or $-\tilde{B}_n$. The decision is the codeword with the smallest metric $M(c_m)$. This metric can be extended to larger codes. Unfortunately, it then soon becomes prohibitively complex. That is, it only works properly if we keep track of the possible transmitted sequences. Clearly, with our system and on a HF channel the DPSK loss over PSK will be considerably smaller than 10dB. To be able to investigate this loss in detail, also the synchronization has to be taken into account. However, the discussion in this section suggests that it might be worthwhile to try to improve the DPSK detector, or use PSK instead.

DISCUSSIONS

We have proposed a wideband OFDM system for the HF channel. It uses DBPSK and powerful coding to handle noise and distortion. To really benefit from a guard-time, a frequency equalizer, to cancel the Doppler spread, is required. For the HF channel, and since the system is very wide-banded, the guard-time can be removed. This is important if we want a military stealth system. Then a guard-time would simplify hostile detection of the transmitted signal.

Table 1: Channel parameters.

Ch.No.	i	a_i	ϕ_i(deg.)	τ_i(ms)	$f_{Di}(t)$(Hz)	ν	f_C(MHz)	Date
0	1	1.00	0.0	0.0	0.0	1.00		
1	1	0.36	249.6	0.82	-0.056	0.72	2.83	23/10/95
	2	0.50	86.5	1.46	-1.185			
	3	1.00	83.4	2.42	-0.696			
2	1	1.00	137.6	1.05	0.156	0.73	4.70	23/10/95
	2	0.36	173.9	3.49	-7.81			
	3	0.33	300.3	5.17	1.09			
	4	0.36	177.3	6.22	0.560			
3	1	1.00	271.0	3.38	-16.1	0.44	21.88	23/10/95
	2	0.78	111.7	4.95	-8.54			
	3	0.81	235.2	5.58	-17.9			
4	1	1.00	154.7	1.41	1.25	0.51	6.78	02/05/96
	2	0.24	134.0	2.81	-0.81			
	3	0.31	232.5	6.40	1.09			
	4	0.36	358.3	8.15	0.56			
	5	0.71	348.5	9.48	3.08			
	6	0.40	205.6	11.60	0.34			

APPENDIX: MULTIPATH PARAMETERS

Based on measurements with DAMSON[1] sounders (Doppler And Multipath SOunding Network), we define 4 channels [5]. The transmitter was located at Harstad, Norway (68.80N 16.50E) and the receiver at Kiruna, Sweden (67.84N 20.40E). The multipath channel is modeled by at most six paths, see Table 1. The phase is selected randomly and the center frequency is denoted by f_C.

References

[1] J.G. Proakis, *Digital Communications*, McGraw-Hill, Third Edition, 1995.

[2] F.J. MacWilliams, N.J.A. Sloane, *The Theory of Error-Correcting Codes*, North-Holland, New York, 1981.

[3] T.C. Giles, *A High-Speed Modem with Built-in HF Noise Tolerance,* Proc of HF Radio Systems and Techniques, York, UK 4-7 July. 1994.

[4] J.E.M Nilsson, *Coding of Spread Spectrum Signals for HF Communications,* Proc. IEEE MILCOM'95, pp.352-356, San Diego, Nov, 1995.

[5] M. Bröms, B. Lundborg, V Jodalen, T Bergsvik, *Doppler Effects on High Latitude HF paths During an Ionospheric Disturbance.* Proc. IEE Seventh International Conf.: HF Radio Systems and Techniques, Nottingham, UK, July 1997.

[6] J.F. Mastrangelo, J.J. Lemmon, L.E. Vogler, J.A. Hoffmeyer, L.E. Pratt, C.J. Behm, *A New Wideband High Frequency Channel Simulation System,* IEEE Trans. Commun., vol. COM-45, pp. 26-34, January 1981.

[1]DAMSON is a collaboration of the Defence Research Agency (DRA), the Norwegian Defence Research Establishment (FFI), the Swedish Defence Research Establishment (FOA) and the Canadian Communications Research Centre (CRC).

EQUALIZATION AND CODING FOR EXTENDED MC-CDMA OVER TIME AND FREQUENCY SELECTIVE CHANNELS

Jochem Egle, Markus Reinhardt, and Jürgen Lindner
Department of Information Technology University of Ulm
Albert Einstein Allee 43 D-89081 Ulm

e-mail: egle@it.e-technik.uni-ulm.de

Abstract

In this work Multi Carrier Code Division Multiple Access (MC-CDMA) – based on Orthogonal Frequency Division Multiplex (OFDM) combined with Hadamard spreading sequences – is extended by an additional spreading in time direction. Introducing the Kronecker Product the two transformations in time and frequency direction can be merged, resulting in one larger transformation.
Furthermore, a suitable equalization algorithm named Iterative Soft Block Decision Feedback Equalizer (ISBDFE) is introduced. The transmission scheme of extended MC-CDMA combined with the equalizer has the capability to transform a time and frequency selective channel asymptotically into a set of AWGN channels. Therefore it is possible to combine this system with Trellis Coded Modulation (TCM) using the codes of Ungerboeck which were originally developed for the AWGN channel. As a lower bound for the uncoded extended MC-CDMA system the Matched Filter Bound (MFB) is introduced. This bound is

K. Fazel and G.P. Fettweis (eds.), Multi-Carrier Spread-Spectrum, 127-134.

equal to the performance of a receiver which has the ability to remove all Inter Symbol Interference (ISI).

1 Introduction

The starting point of this work were the publications of [Rei-95] and [Faz-93]. The work [Rei-95] demonstrates the possibility of transforming a Rayleigh channel into a set of parallel AWGN channels using Hadamard spreading sequences and the ISBDFE equalization algorithm. The second publication introduces MC-CDMA and its properties. As it will be shown in this work the results of these two publications can be easily combined resulting in a new powerful method to cope with time and frequency selective channels. In addition to this the transformation scheme in [Faz-93] is extended by a further transformation in time direction.

2 Transmission with cyclic signals over frequency selective channels – OFDM

OFDM was introduced in the 70s in [Wei-71] and is now again subject of several publications (e.g. [Kor-92]). In OFDM a blockwise transmission of blocklength N_c is assumed, where the transmission blocks are generated by a cyclic extension of each block. If the length of this expansion is L with $L \geq L_0 - 1$ and L_0 the length of the channel impulse response, there will be no Inter Block Interference (IBI). It can be shown for OFDM, if in addition to the cyclic expansion the transmission function is (nearly) constant during one block, that the different subchannels are decoupled (see e.g. [Kor-92]). The inverse Fourier transform in the transmitter and the Fourier transform in the receiver transform a frequency selective channel into N_c parallel AWGN channels with complex amplitude λ_n. The complex amplitudes λ_n of the subchannels are derived from the channel impulse matrix $\boldsymbol{H}$. Note that $\boldsymbol{H}$ is a Toeplitz matrix derived from the channel impulse response $\boldsymbol{h}$. $\boldsymbol{A}^H$ is the conjugate complex transposed of matrix $\boldsymbol{A}$

$$\boldsymbol{D}_\lambda = \boldsymbol{F}_F \boldsymbol{H} \boldsymbol{F}_F^H \tag{1}$$

$$\boldsymbol{D}_\lambda = \mathrm{diag}(\lambda_0, \lambda_1, \ldots, \lambda_{N_c-1}) \text{ with} \tag{2}$$

$$\boldsymbol{\lambda} = \sqrt{N_c} \boldsymbol{F}_F \cdot \boldsymbol{h}\,. \tag{3}$$

3 Transmission scheme

In this section a transmission scheme for OFDM is introduced. The problem of OFDM are the different channel amplitudes of the subchannels. These amplitudes are correlated and Rayleigh distribued. Therefore the transmission characteristic of one subchannel as introduced in the previous section is that of a flat Rayleigh fading channel. Due to this fact diversity techniques such as frequency and time diversity are necessary to improve the transmission performance. This is achieved by spreading the transmission symbols by Hadamard sequences in time and frequency direction. The transmission model is shown in figure 1. The sequences are

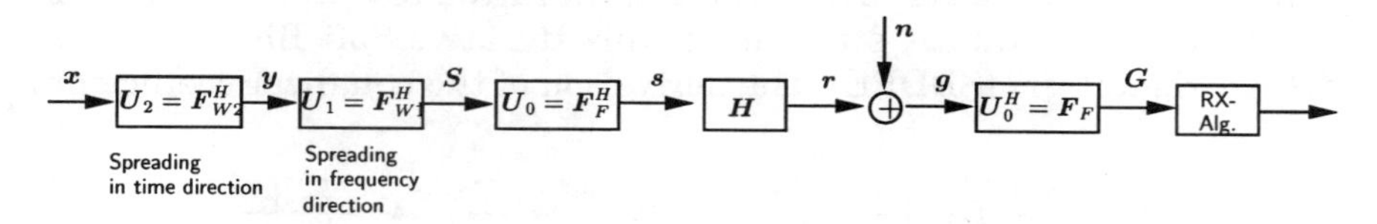

Figure 1: Model of a extended MC-CDMA; $\boldsymbol{F}_{W1}$ (length N_F) denotes the spreading in frequency direction and $\boldsymbol{F}_{W2}$ (length N_T) in time direction

orthonormal. In practical implementations fast algorithms are used. In the following it will be shown that for combined spreading in time (over the OFDM blocks) and frequency direction (over the subchannels) the resulting transformation $\boldsymbol{F}_W$ is again a Hadamard transformation. The spreading in time direction can be written as:

$$\bar{\boldsymbol{Y}} = [\boldsymbol{y}(1)\,\boldsymbol{y}(2)\,\ldots\,\boldsymbol{y}(N_T)] = [\boldsymbol{x}(1)\,\boldsymbol{x}(2)\,\ldots\,\boldsymbol{x}(N_T)] \cdot \boldsymbol{F}_{W2}^H \quad ,$$

or with the vec operator and the Kronnecker product $\otimes$

$$\widetilde{\boldsymbol{a}} = \mathrm{vec}\,\bar{\boldsymbol{A}} = [\boldsymbol{a}(1)^T\,\boldsymbol{a}(2)^T\,\ldots\,\boldsymbol{a}(N)^T]^T$$

$$\mathrm{vec}\,\bar{\boldsymbol{Y}} = \widetilde{\boldsymbol{y}} = \widetilde{\boldsymbol{F}}_{W2}^H \cdot \widetilde{\boldsymbol{x}} \quad \text{with} \quad \widetilde{\boldsymbol{F}}_{W2} = \boldsymbol{F}_{W2} \otimes \boldsymbol{I}_{N_F \times N_F} \quad .$$

The additional spreading in frequency direction can be written as

$$\widetilde{\boldsymbol{S}} = \boldsymbol{I}_{N_T \times N_T} \otimes \boldsymbol{F}_{W1}^H \cdot \widetilde{\boldsymbol{y}} \quad .$$

The resulting transformation can be written as

$$\boldsymbol{F}_w = \widetilde{\boldsymbol{F}}_{W2} \cdot \widetilde{\boldsymbol{F}}_{W1} = (\boldsymbol{F}_{W2} \otimes \boldsymbol{I}_{N_F \times N_F}) \cdot (\boldsymbol{I}_{N_T \times N_T} \otimes \boldsymbol{F}_{W1}) = \boldsymbol{F}_{W2} \otimes \boldsymbol{F}_{W1} \,. \tag{4}$$

Another possibility is not to spread over all subchannels and to perform an additional interleaving to minimize correlation between different subchannels used by one spreading block. This results in a decrease of complexity. Furthermore, it is possible to increase performance or to decrease complexity by performing a block interleaving in time direction. The best solution depends on the coherence time and frequency of the channel. A powerful tool to estimate the performance is the matched filter bound derived in section 5.

4 Iterative Soft Block Decision Feedback Equalizer

The transmission over the OFDM system destroys the orthogonality of the Hadamard sequences. Therefore suitable equalization algorithms have to be applied. Equalizers based on maximum likelihood are optimum, but the complexity grows exponentially. Therefore a low complexity iterative equalizer based on the soft decision feedback introduced in [Rei-95] is applied. Figure 2 shows the structure of this Iterative Soft Block Decision Feedback Equalizer (ISBDFE). The derivation of the equalizer coefficients

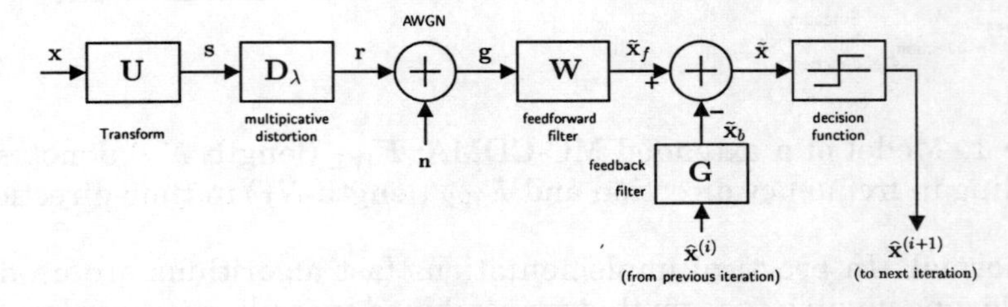

Figure 2: Iterative Soft Block Decision Feedback Equalizer

(matrices $\boldsymbol{W}$ and $\boldsymbol{G}$) is based on Minimum Mean Square Error (MMSE) criterion

$$\mathcal{J} = \mathrm{E}\left\{\|\boldsymbol{x} - \tilde{\boldsymbol{x}}^{(i)}\|^2\right\} \stackrel{!}{=} \min_{\boldsymbol{W},\boldsymbol{G}} \quad .$$

With the help of this criterion we derive after several algebraic manipulations the following results.
For the forward filter we get

$$\begin{aligned} \boldsymbol{W} &= \boldsymbol{U}^H \cdot \boldsymbol{D}_{\tilde{\lambda}} \\ \text{with } \tilde{\lambda}_i &= \frac{1}{|\lambda_i|^2(1-\varrho^2)+N_0} \cdot \frac{\lambda_i^*}{1+\frac{\varrho^2}{N_t}\cdot\sum\limits_{i=1}^{N_t}\frac{|\lambda_i|^2}{|\lambda_i|^2(1-\varrho^2)+N_0}}, \end{aligned} \tag{5}$$

and for the feedback filter matrix

$$\boldsymbol{G} = \varrho\left(\boldsymbol{U}^H\,\boldsymbol{D}_{\lambda\tilde{\lambda}}\,\boldsymbol{U} - \boldsymbol{I}\frac{1}{N_t}\mathrm{trace}\left\{\boldsymbol{D}_{\lambda\tilde{\lambda}}\right\}/N\right) \text{ with } \boldsymbol{D}_{\lambda\tilde{\lambda}} = \boldsymbol{D}_\lambda\boldsymbol{D}_{\tilde{\lambda}} \quad . \tag{6}$$

In these equations the factor $\varrho \in [0,1]$ is a measurement for the correlation between transmitted and received symbols. Therefore we call this type of equalizer **soft** block decision feedback equalizer. i.e. for high SNR ϱ is approximately 1. For the initialization of the feedback a linear frontend is used, which means that ϱ is set to 0. Practical simulations with the TUX (GSM channel model for a typical urban environnenment) have shown.

that $\varrho = 1 - P_{S_{RAY}}$ is a good choice. In this equation $P_{S_{RAY}}$ denotes the symbol error rate for the Rayleigh channel.
Note furthermore that for extended MC-CDMA $\boldsymbol{U}$ is substituted by $\boldsymbol{F}_W$.

5 Matched Filter Bound

With the help of the MFB (see also [Rei-97] and [Fec-93]) it is possible to calculate a lower bound for the BER of the extended MC-CDMA system. For the calculation it is assumed, that the deorthogonalization is completely removed.
We assume for the derivation without loss of generality that the first element $x(1) = 1$ of the vector $\boldsymbol{x}$ is transmitted over the OFDM channel. This symbol is spread by the first row of the matrix $\boldsymbol{F}_W^H$ (denoted by $\boldsymbol{f}_{W1}^H$) and transmitted over the channel. At the output of the filter we get

$$x(1) = \boldsymbol{f}_{W1}\boldsymbol{D}_\lambda^*\boldsymbol{D}_\lambda\boldsymbol{f}_{W1}^H = \frac{1}{N_T \cdot N_F}\boldsymbol{\lambda}^H\boldsymbol{\lambda} = \frac{1}{N}\boldsymbol{\lambda}^H\boldsymbol{\lambda} \quad \text{with} \quad \boldsymbol{\lambda} = \text{diag}(\boldsymbol{D}_\lambda) \; . \quad (7)$$

For the BER (2PSK) of the transmission we can derive

$$P_b = \int_0^\infty P_b(x) \cdot p(x)dx = \int_0^\infty \frac{1}{2}\text{erfc}\left(\sqrt{\frac{1}{2} \cdot x\frac{E_b}{N_0}}\right) \cdot p(x)dx \; , \quad (8)$$

where $p(x)$ denotes the Probability Density Function (PDF) of the symbol $x(1)$ at the matched filter output. The PDF will be calculated with the help of the characteristic function

$$\Psi(j\nu) = \int_{-\infty}^{+\infty} \exp(j\nu x)p(x)dx = \text{E}\{\exp(j\nu x(1))\} = \text{E}\{\exp(j\nu\frac{1}{N}\boldsymbol{\lambda}^H\boldsymbol{\lambda})\} \; .$$

Now we assume according to the channel model that $\boldsymbol{\lambda}$ is a vector of correlated gaussian random variables derived from the uncorrelated noise $\boldsymbol{n}$

$$\boldsymbol{\lambda}^H\boldsymbol{\lambda} = \boldsymbol{n}^H\boldsymbol{A}^H\boldsymbol{A}\boldsymbol{n} = \boldsymbol{n}^H\boldsymbol{\Phi}_{\lambda\lambda}\boldsymbol{n} \text{ with } \boldsymbol{\lambda} = \boldsymbol{A} \cdot \boldsymbol{n} \; .$$

With the matrix $\boldsymbol{\Phi}_{\lambda\lambda}$ as a non-negative definite Hermitian matrix which can be diagonalized by an orthonormal matrix $\boldsymbol{Q}$ resulting in a diagonal matrix $\boldsymbol{\Gamma}$.
The characteristic function of the random variable $x(1)$ is now

$$\begin{aligned}\Psi(j\nu) &= \text{E}\{\exp(j\nu\boldsymbol{n}^H\boldsymbol{\Gamma}\boldsymbol{n})\} = \text{E}\{\exp(j\nu\sum_{i=1}^{N}\gamma_i \cdot |\tilde{n}_i|^2)\} \\ &= \text{E}\{\prod_{i=1}^{N}\exp(j\nu\gamma_i \cdot |\tilde{n}_i|^2)\} = \prod_{i=1}^{N}\frac{1}{1 - j\nu\gamma_i}\end{aligned}$$

In practical calculations only some (say m) Eigenvalues are significant. The others are nearly zero. To simplify the inverse transform the characteristic function is expressed as a sum

$$\Psi(j\nu) \approx \sum_{i=1}^{m} \frac{m_i}{1 - j\nu\gamma_i} \quad \text{with} \quad m_i = \prod_{\substack{j=1 \\ j\neq i}}^{m} \frac{1}{1 - \gamma_j/\gamma_i}.$$

Thus the PDF is

$$p(x) = \sum_{i=1}^{m} \frac{m_i}{\gamma_i} \exp(-x/\gamma_i). \tag{9}$$

The problem is the huge dimension of the correlation matrix (because of the spreading in time and frequency direction). The size N is the product of the length in time N_T and in frequency N_F direction. Therefore the calculation of the eigenvalues may become a difficult task.
It can be shown (see [Rei-97]) , that if the correlation function is seperable in the arguments in time and frequency direction it is possible to separate the problem in one decomposition in frequency direction (size $N_F \times N_F$) and one in time direction (size $N_T \times N_T$). The diagonal matrix of the whole system can be computed with the Kronecker product of the two diagonal submatrices

$$\mathbf{\Gamma} = \mathbf{\Gamma}_F \otimes \mathbf{\Gamma}_T. \tag{10}$$

6 Combination of extended MC-CDMA with TCM

In order to improve the feedback and to get a coding gain Trellis Coded Modulation (TCM) is applied. Because of the near AWGN characteristic of the uncoded extended MC-CDMA transmission scheme the codes of Ungerboeck can be used. In Figure 3 the receiver is shown. For the initialization similar to the uncoded case a linear frontend is used. Note that the feed-

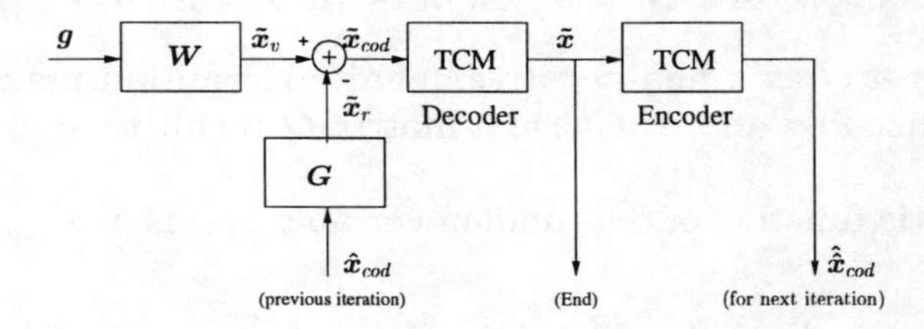

Figure 3: Iterative Block Decision Feedback Equalizer with TCM Coding

back is improved by the decoding of the symbols in the feedback.

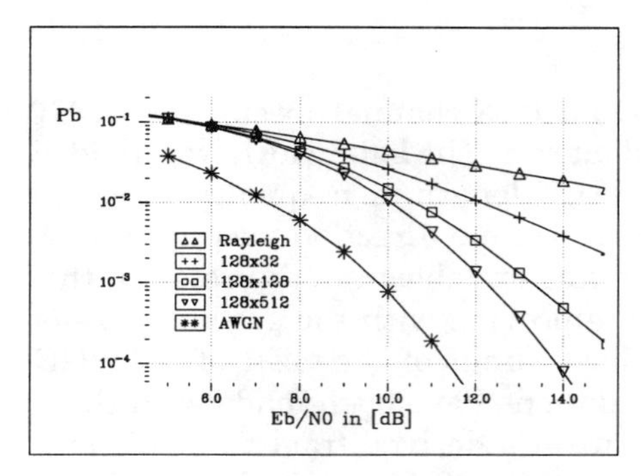

Figure 4: Different sizes of spreading matrices, ϱ calculated with Rayleigh BER, TUX 100 km/h, 3 iterations

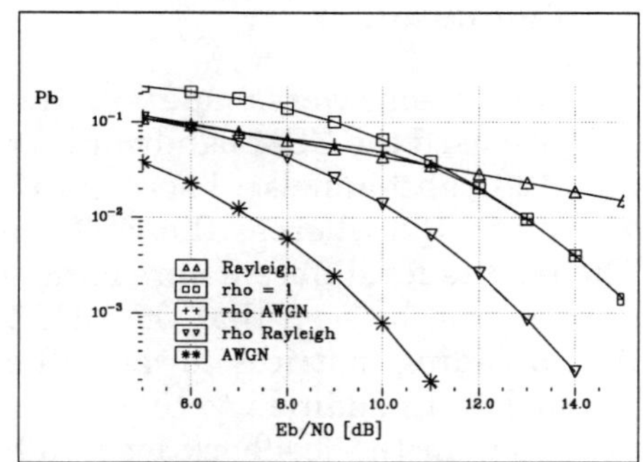

Figure 5: Effects of the Soft Feedback, comparison of $\varrho = 1$, calculated with AWGN and with Rayleigh BER, TUX 100 km/h, spreading length 128x256, 3 iterations

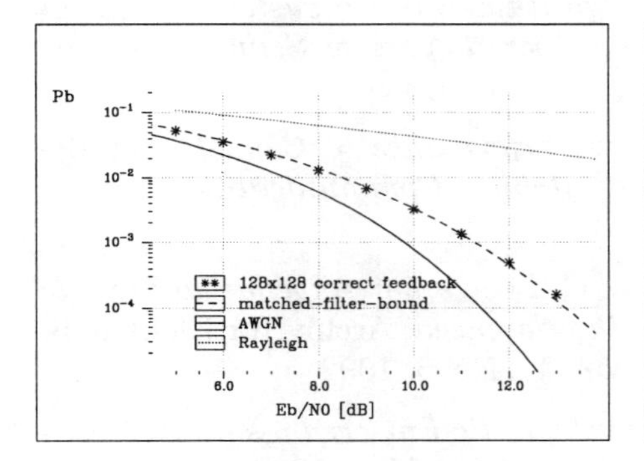

Figure 6: Verification of the Matched Filter Bound, comparison between calculation and simulation with perfect equalization, TUX 100 km/h, spreading 128x128

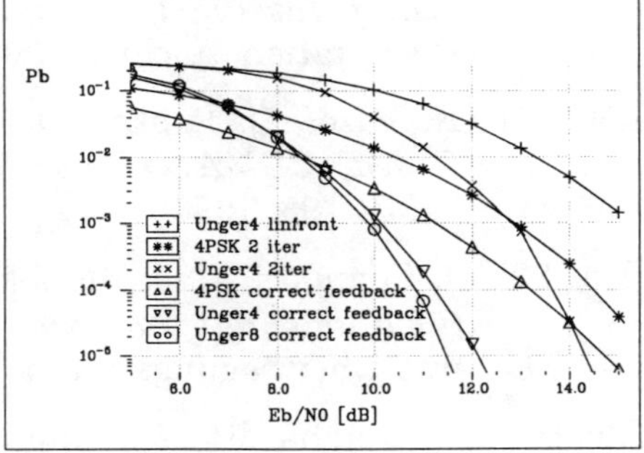

Figure 7: combined Trellis Coded Modulation Unger4 denotes the code with memory 4 whereas Unger8 with memory 8, TUX 100 km/h spreading 128x256

7 Simulations

The simulations were made with the GSM TUX channel specified in COST 207. We used the GSM bandwidth and bitrate. The bandwidth was divided into 128 subchannels. Therefore the FHT length in frequency direction was up to 128 whereas the FHT length in time direction was up to 512. The results for different spreading lengths are shown in figure 4. In this figure it can be seen that the BER is improving with the growing size of the spreading matrices towards the lower limit of the BER of a AWGN channel. In addition to this all simulations have been made with three iterations in the feedback loop and with a ϱ derived from the BER of a Rayleigh channel. In figure 5 one can see the effects of the soft feedback in the receiver structure. The best performance was achieved with a ϱ derived from the BER of a Rayleigh channel. In figure 6 a proof for the validity of the MFB is given. This chart shows the comparison between calculated and simulated BER for the GSM TUX. For TCM the codes of Ungerboeck were used. In figure 7 it can be seen that for higher SNR (13 dB) TCM provides a significant improvement whereas for lower SNR the uncoded transmission is better.

References

[Fec-93] S. Fechtel:*Verfahren und Algorithmen der robusten Synchronisation für die Übertragung über dispersive Schwundkanäle*, Dissertation, Aachen, 1993

[Faz-93] K. Fazel, L. Papke: *On the Performance of Convolutionally-Coded CDMA/OFDM for Mobile Communication System* PIMRC'93, 1993

[Kor-92] J. Benndorf, D. J. Korobkov, G. Schwanja: *Angepasste Signale für Kanäle mit Intersymbolinterferenzen*, Archiv für Elektronik und Übertragungstechnik, Vol. 46, No. 6, 1992

[Rei-97] M. Reinhardt: *Kombinierte vektorielle Entzerrungs- und Decodierverfahren*, Dissertation (eingereicht), Ulm, 1997

[Rei-95] M. Reinhardt, J. Lindner: *Transformation of a Rayleigh fading channel into a set of parallel AWGN channels and its advantage for coded transmission*, Electronics Letters, vol. 31, no. 25, pp. 2154–2155, 7th Dec. 1995

[Wei-71] S. B. Weinstein, P. M. Ebert: *Data Transmission by Frequency–Division Multiplexing Using the Discrete Fourier Transform*, IEEE Trans. Commun. Tech., Vol. COM–19, pp. 628–634, Oct. 1971

DETECTION METHOD FOR MC-CDMA BASED ON A RECURRENT NEURAL NETWORK STRUCTURE

W.G. Teich, J. Egle, M. Reinhardt, and J. Lindner

Department of Information Technology, University of Ulm
e-mail: teich@it.e-technik.uni-ulm.de

Abstract

A detection method based on a recurrent neural network structure is derived for a multi carrier code division multiple access communication system with multi path propagation. Contrary to other neural network approaches, the RNN has the advantage, that network size as well as the coefficients of the network can be derived from parameters which characterize the communication system. The energy function of the RNN is identical to the log-likelihood function of the maximum likelihood detector. Different iteration algorithms for the RNN with an emphasis on parallel processing are discussed. Performance results are given for the Rayleigh fading channel and a typical mobile radio channel. Performance and complexity of the RNN detector are compared with other iterative detection algorithms, specifically a block decision feedback equalizer.

1 INTRODUCTION

Multi carrier code division multiple access (MC-CDMA) is a promising technique for future mobile communication systems [1, 2, 3, 4]. MC-CDMA allows to combine the advantages of CDMA systems (flexibility, multi path immunity)

K. Fazel and G.P. Fettweis (eds.), Multi-Carrier Spread-Spectrum, 135-142.

and orthogonal frequency division muliplexing (OFDM) techniques (low receiver complexity).

The general idea of MC-CDMA is to transmit an information symbol not only over a single carrier (which might be subject to strong fading), but to spread the symbol over several (independently fading) carriers. In this way a diversity gain can be achieved and the system performance improved. For time variant multi path channels, the diversity gain can be further increased by additionally spreading the information symbol in time domain. In order to allow an efficient implementation, the spreading in time and frequency is realized using orthogonal Walsh-Hadamard transformations. After spreading, the symbols are transmitted by a conventional OFDM scheme with guard time. The flexibility of the system can be achieved by different assignments between users and Walsh-Hadamard codes. At the receiver side, a channel matched filter followed by a detection algorithm is used. A more detailled description of the transmission scheme is given in [5].

Due to multi path propagation, the Walsh-Hadamard codes are deorthogonalized. This leads to intersubchannel interference. Thus an equalizer is required at the receiver side. Note, that, in contrast to direct sequence (DS-) CDMA systems, the complexity of the equalizer can be adjusted by the choice of the length of the spreading codes.

The equalizer operates on the complete spreading block simultaneously. In [5] an iterative detection algorithm, the block decision feedback equalizer with soft feedback (ISBDFE) is described. Motivated by results obtained for DS-CDMA systems [6] we investigated an equalizer based on a recurrent neural network (RNN) structure [7]. In principle, also other neural networks (NN) can be used as detection algorithm. However, compared to, e.g., a multilayer perceptron, the RNN has various advantages, mainly, that size and parameters of the RNN can be obtained from the parameters of the respective communication system.

2 TRANSMISSION MODEL FOR MC-CDMA

Lindner gives the general model for a continuous and discrete time vector transmission (cf. Fig. 3 in [4]). Depending on the assignment between different subchannels and users, Fig. 1 gives either the model for MC-CDMA or multi subchannel transmission. The base is a blockwise transmission with OFDM. By cyclic extension (guard time) all interblock interference (IBI) is suppressed. By spreading all information symbols in frequency direction a diversity gain can be achieved. For complexity reasons a Fast-Hadamard-Transformation (FHT) is used for spreading. Transmitting over frequency selective channels leads to a deorthogonalization and subsequently to multiple access interference (MAI)

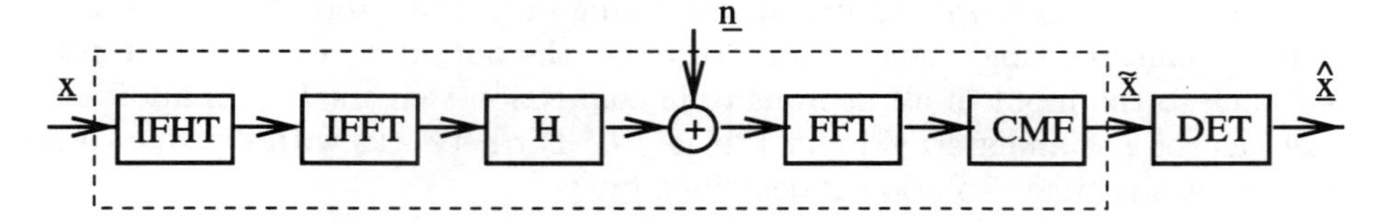

Figure 1: MC-CDMA: Discrete time vector transmission model. FHT: Fast-Hadamard-Transformation; FFT: Fast-Fourier-Transformation; Ixxx denotes the respective inverse transformations. The detection algorithm "DET" operates on the output vector $\tilde{\underline{x}}$ of the channel matched filter (CMF).

and/or intersymbol interference (ISI):

$$\tilde{x}_i = R_{ii}x_i + \sum_{j=1(j\neq i)}^{N} R_{ij}x_j + n_i.$$

Here, x_i are the transmitted symbols, $\tilde{x}_i$ the symbols obtained at the output of the channel matched filter (CMF), and n_i are the components of the noise vector at the output of the CMF. R_{ij} are the elements of the discrete time channel matrix $\boldsymbol{R}$.

3 DETECTION ALGORITHMS

The maximum liklihood (ML) detection algorithm gives optimal performance. The output of the ML detector is the estimated vector $\hat{\underline{x}}$ which maximizes the log-likelihood function. For BPSK modulation the log-likelihood function can be derived to be:

$$\Lambda = \sum_{i=1}^{N} \hat{x}_i\tilde{x}_i - \frac{1}{2}\sum_{i=1}^{N}\sum_{j=1}^{N} R_{ij}\hat{x}_i\hat{x}_j. \tag{1}$$

However, due to the large complexity (growing exponentially with the spreading length n_s) of the ML detector, suboptimal detection algorithm are needed. The vector transversal equalizer (TE) combines low complexity with moderate performance. Using the ISBDFE the performance can be substantially improved, while the complexity still remains moderate [5]. Note, that for block transmission it is natural to iterate the decision feedback. In this paper we introduce an iterative detector based on a RNN structure.

3.1 RNN detector

In the following we give a short introduction to RNN. For a detailed discussion of neural networks we refer to the literature (e.g., [8]). A RNN consists of a large number N of simple nonlinear processing units ("neurons"). It is usually a fully connected single-layered structure, i.e. the output V_j of each neuron is fed back to the input of all neurons i via connection weights T_{ij}. In addition, each neuron has an external input ("bias") I_i. For networks with discrete time dynamics, the time evolution is described by

$$V_i(l+1) = \sigma\left(\sum_{j=1}^{N} T_{ij}V_j(l) + I_i\right). \tag{2}$$

As nonlinear activation function σ we use a hyperbolic tangen function (see also [9]), i.e. $\sigma(x) = \tanh(c \cdot x)$. Here, c is a positive constant which controls the slope of the activation function for $x = 0$, i.e. it defines the width of the transition region. For $c \to \infty$ the sign function is recovered. The slope c is a free parameter of the RNN.

Motivated by analogies to physical systems (spin glasses) one can define the energy function

$$E = -\sum_{i=1}^{N} V_i I_i - \frac{1}{2}\sum_{i=1}^{N}\sum_{j=1}^{N} T_{ij}V_iV_j \tag{3}$$

for the RNN. For a symmetric connection weight matrix $T_{ij} = T_{ji}$, Hopfield and Tank [7] have shown, that the system always converges to a stable state (fixed point) for a sequential updating procedure.

Comparing the log-likelihood function (1) of the ML detector with the energy function of a RNN (3), we notice a strong correspondence. Choosing one neuron for each element of the transmission vector, $I_i = \tilde{x}_i$, $T_{ij} = -R_{ij}$ and identifying the output of the network with the estimated transmission vector $\hat{\underline{x}}$, we obtain $E = -\Lambda$. This implies, that the global minimum of the energy function of the RNN corresponds exactly to the global maximum of the likelihood function. The problem that remains is to find a time evolution of the network (an iteration process in the case of a discrete time network) which assures that the global minimum of the energy function is found, or at least a state which does not deviate much from the global minimum.

For $\hat{x}_i \in \pm 1$ the diagonal terms ($i = j$) in the second term of the log-likelihood function (1) can be neglected, since they give a constant contribution only. For the RNN detector this implies, that the self-connections vanish ($T_{ii} = 0$). For suitable values of the slope c of the activation function, it can then be shown, that the outputs of the neurons converge to ± 1 [9]. The structure of the RNN detector then has a very simple and intuitive interpretation. Consider a neuron and assume that the output of all neurons connected to the input of this neuron i are correct, i.e. $V_j = x_j$. In this case all interference (ISI and MAI) present in the external input $\tilde{x}_i$ of neuron i is completely cancelled by the feedback terms.

In order to find a minimum of the energy function, various iteration algorithm can be applied. Keeping the slope c constant during the iteration process, the optimum slope can be found by a training procedure minimizing the bit error rate (BER) [6]. For a constant slope $c \to \infty$ the RNN detector is identical to an iterative block decision feedback equalizer with hard feedback [5]. However, for strong interference channels and c =const. the iteration process ends in local minima, resulting in a poor performance. To improve performance, a process similar to simulated annealing has been applied. We start the iteration process with a slope $c = 0$ and gradually increase it up to the maximum value c_{max}: $c_l = \frac{c_{max}}{i_{max}} l$. The maximum number of iterations i_{max} is kept fixed, while c_{max} is determined by minimizing the BER.

Two fundamental update procedures can be distinguished: sequential updating and parallel updating. For sequential updating the iteration process always reaches a fixed point. For parallel update, however, the output of the neurons can oscillate, resulting in a bad performance. In order to avoid these oscillations and still utilizing as much parallelism as possible, we applied an update procedure which is parallel in groups. Between different groups the updating is sequential. The partitioning of the neurons into groups should be performed in such a way as to minimize the coupling between the groups.

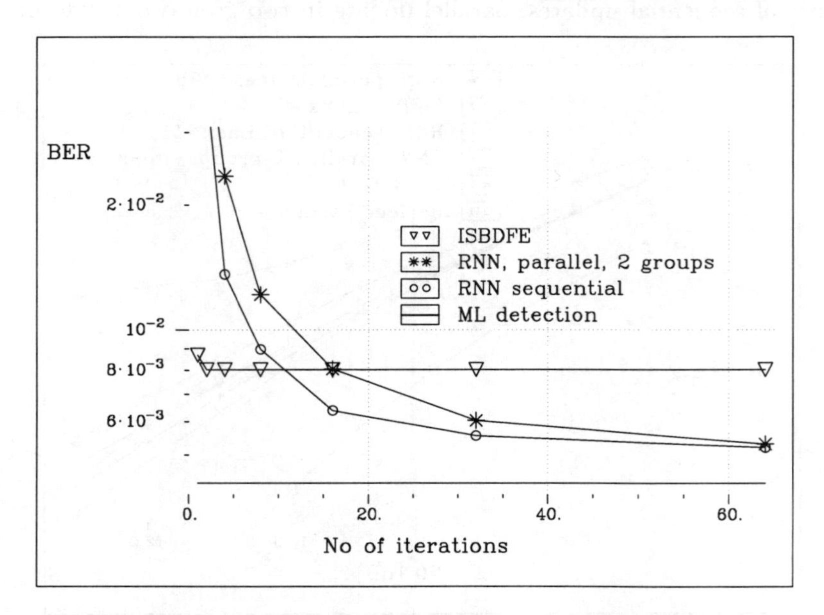

Figure 2: BER as a function of the number of iteration steps per neuron. Parameters: Perfect interleaved Rayleigh channel, BPSK modulation, spreading length $n_s = 8$, $E_b/N_0 = 12$ dB, linear increasing slope c of activation function.

4 SIMULATION RESULTS

Simulations for different mobile radio channels have been performed. We used BPSK modulation and assumed perfect synchronization between transmitter and receiver and perfect knowledge of the channel parameters in the receiver.

Fig. 2 shows the BER as a function of the number of iteration steps l. For the ISBDFE no further improvements can be achieved after two or three iterations. In contrast, the RNN detector asymptotically approaches ML performance with increasing number of iteration steps. For the RNN detector, sequential update converges faster than parallel update in two groups. Pure parallel update is not shown in Fig. 2, but has a much poorer performance and does not reach ML performance asymptotically.

Fig. 3 shows the BER as a function of E_b/N_0 for the perfectly interleaved Rayleigh channel. In this case the optimal diversity gain can be achieved (curve with perfect feedback in Fig. 3). Fig. 4 shows the results for a time variant and frequency selective mobile radio channel (TUX). For both channels pure parallel update has a very bad performance due to oscillations in the output vector. However, for sequential update or for parallel update in two groups, the performance of the ISBDFE can be improved by up to 1 dB. For the same number of sequential updates, parallel update in two groups outperforms se-

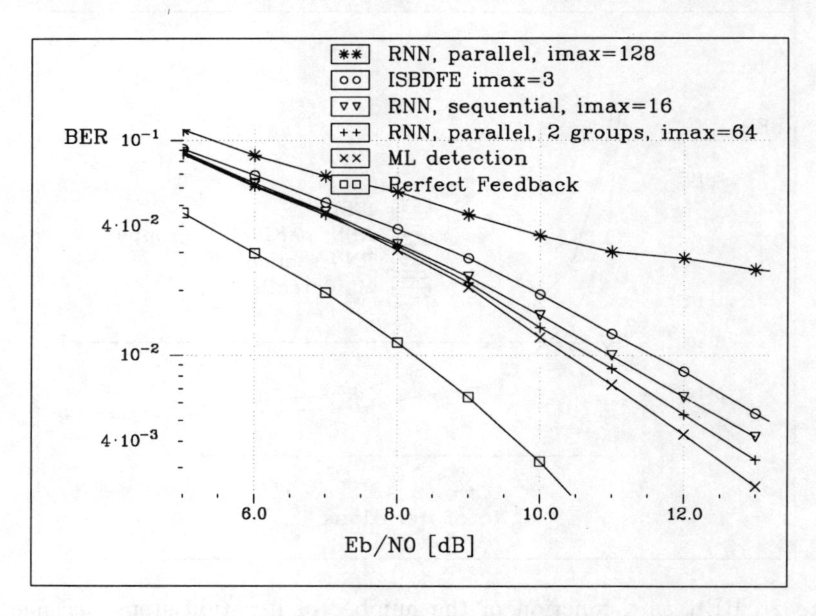

Figure 3: BER as a function of E_b/N_0 for the perfect interleaved Rayleigh channel. Parameters: spreading length $n_s = 8$, BPSK modulation, linear increasing slope c of activation function. The two-sided noise power spectral density is used.

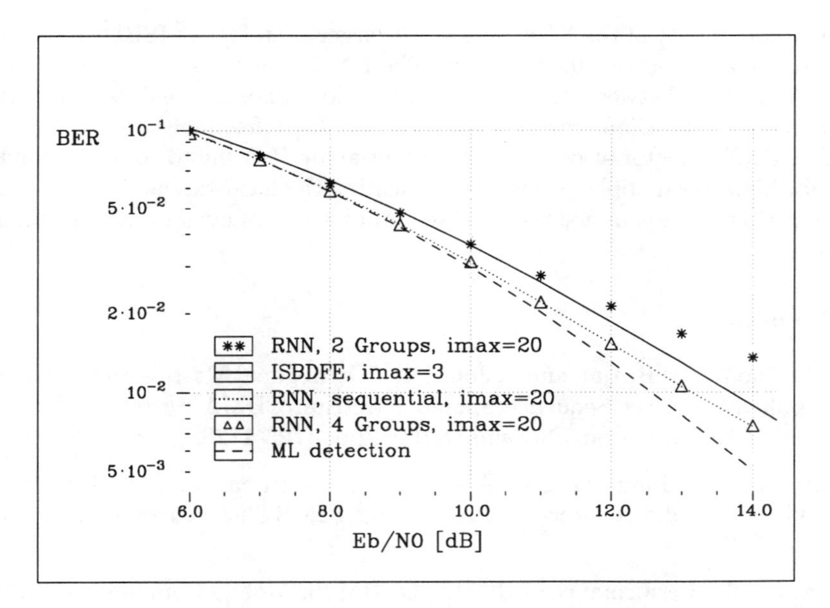

Figure 4: BER as a function of E_b/N_0 for the TU100 channel. Spreading and interleaving in frequency domain: $2x64$ carriers, in time domain: $4x16$ blocks. Parameters: BPSK modulation, interpolation function rect, symbol duration: $T_s = 0.47$ msec, linear increasing slope c of activation function. The two-sided spectral noise power density is used.

quential updating. For low E_b/N_0 the BER is dominated by additive noise and ML performance is reached with few iterations. Increasing E_b/N_0 the interference terms become more important and increasing the number of iterations can further reduce these interference terms and consequently improve the BER.

Similar results can be obtained for a time invariant and frequency selctive twopath channel.

5 SUMMARY

We have shown, that a RNN detector can be designed in such a way, that the energy function of the RNN resembles the log-likelihood function of the underlying detection problem. By using simulated annealing and parallel update in groups improvements of up to 1 dB compared with the ISBDFE can be obtained. Increasing the number of iterations, ML performance is approached asymptotically.

Compared to other neural network approaches, the RNN detector has only a minimal training complexity. Network size and connectivity are easily derived from the system parameters.

The complexity of the RNN detector is larger than for the ISBDFE, but still moderate compared to ML detection. The RNN detector shows a large flexibility. A tradeoff between complexity and performance is possible. Increasing the number of iterations results in a better BER performance.

The RNN detector also has a high potential for distributed computation and thus for high speed applications. By replicating the hardware and transforming the iteration process in a stage by stage processing, an even larger throughput is possible.

References

[1] A. Chouly, A. Brajal, and J. Jourdan, "Orthogonal Multicarrier Technique applied to Direct Sequence Spread Spectrum CDMA Systems", Proceedings IEEE Globecom, November 1993, pp. 1723-1728.

[2] N. Yee, J.P. Linnartz, and G. Fettweiss, "Multi carrier CDMA in Indoor Wireless Radio Networks", Proceedings PIMRC'93, Yokohama, pp. 109-113, 1993.

[3] K. Fazel, "Performance of CDMA/OFDM for Mobile Communication Systems", Proceedings ICUP'93, pp. 975-979, 1993.

[4] J. Lindner, "MC-CDMA and General Multiuser/Multisubchannel Transmission Techniques - An Overview", IEEE Multi-Carrier Spread-Spectrum Workshop, Oberpfaffenhofen, April 24-25, 1997 (invited paper).

[5] J. Egle, M. Reinhardt, and J. Lindner, "Equalization and Coding for extended MC-CDMA over Time and Frequency Selective Channels", IEEE Multi-Carrier Spread-Spectrum Workshop, Oberpfaffenhofen, April 24-25, 1997.

[6] W.G. Teich and M. Seidl, "Code Division Multiple Access Communications: Multiuser Detection based on a Recurrent Neural Network Structure", Proceedings IEEE ISSSTA'96, September 22-25, 1996, Mainz, Germany, pp. 979-984.

[7] J.J. Hopfield, D.W. Tank, " "Neural" Computation of Decisions in Optimization Problems", Biological Cybernetics, vol. 52, pp. 141–152, 1985.

[8] D.R. Hush and B.G. Horne, "Progress in Supervised Neural Networks - What's New Since Lippmann?", IEEE Signal Processing Magazine vol. 1, pp. 8-39, January 1993.

[9] T. Miyajima, T. Hasegawa, M. Haneishi, "On the Multiuser Detection Using a Neural Network in Code–Division Multiple–Access Communications", IEICE Transactions on Communications, vol. E76–B, pp. 961–968, August 1993.

Section IV

SYNCHRONIZATION AND CHANNEL ESTIMATION

SENSITIVITY OF OFDM/CDMA TO CARRIER PHASE JITTER

Heidi Steendam, Marc Moeneclaey

Communications Engineering Lab.
University of Ghent
B-9000 GENT, BELGIUM

ABSTRACT

In this paper, we investigate the influence of carrier phase jitter on the performance of an OFDM/CDMA system. This carrier phase jitter reduces the performance of the system, defined as the signal-to-noise ratio at the input of the decision device. When all users have the same power level and phase jitter spectrum, it is shown that for the highest load, the degradation only depends on the jitter variance but not on the specific shape of the jitter spectrum.

INTRODUCTION

Orthogonal frequency-division multiplexing code-division multiple-access (OFDM/CDMA) has recently received a considerable attention owing to its possibility to achieve a high capacity per unit bandwidth, which provides us an excellent candidate for high data rate applications [1]-[6].

K. Fazel and G.P. Fettweis (eds.), Multi-Carrier Spread-Spectrum, 145-152.

OFDM/CDMA consists of a combination of two well known and studied modulation techniques. In CDMA, the different users are modulated by codes. All users can transmit at the same time and, as each user makes use of the entire frequency band, a high capacity of the channel is reached. A drawback of CDMA is that the performance rapidly decreases when the number of users increases.
Orthogonal frequency-division multiplexing (OFDM) multiplexes the incoming streams on orthogonal carriers. These carriers are spectrally overlapping, in contrast to FDMA which uses non-overlapping channels. Moreover, OFDM is less complex to realize, as it can be implemented using a fast Fourier transform algorithm (FFT).
It is known that multi-carrier systems are more sensitive to carrier phase jitter than single carrier systems [7]. This carrier phase jitter is generated by the phase locked loop (PLL) which converts the received IF signal to the base-band signal. In the OFDM/CDMA system, the carrier phase jitter occurring in the PLL systems gives rise to the multi-user interference (MUI).
In this paper, we derived the degradation of the OFDM/CDMA system caused by carrier phase jitter. An expression for the degradation is presented in terms of the phase jitter spectra and the power levels.

SYSTEM DESCRIPTION

The conceptual block diagram of the OFDM/CDMA transceiver is shown in Fig. 1. A data symbol $a_{n,m}$ with unit energy and at symbol rate 1/T, transmitted by the nth user during the mth symbol interval, is multiplied with a CDMA chip sequence $\{c_{n,\ell}\}$ $\ell=0,\ldots,N-1$, N denoting the number of chips. Sequences belonging to different users are assumed to be orthogonal. The resulting samples, at a rate N/T, are modulated on N equidistant orthogonal carriers using an inverse discrete Fourier transform. The resulting time domain samples are fed to p(t), a unit energy square root Nyquist filter with respect to the interval T/N. The complex envelope s(t) of the transmitted signal is disturbed by additive noise and carrier phase jitter. This carrier phase jitter is the phase error between the carrier used for up-converting the baseband OFDM/CDMA signal at the transmitter and the phase-locked carrier used for down-converting at the receiver. All OFDM carriers exhibit an identical carrier phase jitter as they are up-converted by the same oscillator. The phase error $\phi_n(t)$ of user n is modeled as a stationary zero mean process having a bandwidth much smaller than N/T. The additive noise n(t) has a power spectrum $S_n(f)$. The complex envelope of the received signal r(t) is given by :

$$r(t)=\sum_{m}\sum_{n=0}^{N-1}\frac{\sqrt{E_{s,n}}}{N}a_{n,m}\sum_{k,\ell=0}^{N-1}c_{n,\ell}e^{j2\pi\frac{k\ell}{N}}p\left(t-(k+mN)\frac{T}{N}\right)e^{j\phi_n(t)}+n(t) \quad (1)$$

The receiver consists of a filter $p^*(-t)$ matched to the transmit pulse whose output is sampled at the chip rate at the instants {kT/N}. In order to detect the symbol $a_{n,m}$, the matched filter output samples are fed to the discrete Fourier transform and the resulting frequency domain samples are correlated with the chip sequence corresponding to the nth user.

SENSITIVITY TO CARRIER PHASE JITTER

In this section, we compute the degradation (in dB) of the signal-to-noise ratio (SNR) at the input of the decision device when carrier phase jitter is present. We consider the detection of the symbol $a_{n,0}$.
As the spectrum of the phase jitter is much smaller than N/T, the variation of the phase jitter over the impulse response duration of the matched filter, which is in the order of T/N, can be neglected. The input of the decision device is fed with the sample $\hat{a}_{n,0}$:

$$\begin{aligned}\hat{a}_{n,0} &= \sqrt{E_{s,n}}\, a_{n,0} \underline{c}_n^T E[B_n] \underline{c}_n^* + \sqrt{E_{s,n}}\, a_{n,0} \underline{c}_n^T \left(B_n - E[B_n]\right) \underline{c}_n^* \\ &+ \sum_{\substack{\ell=0 \\ \ell \neq n}}^{N-1} \sqrt{E_{s,\ell}}\, a_{\ell,0} \underline{c}_\ell^T E[B_n] \underline{c}_n^* + W_{n,0}\end{aligned} \tag{2}$$

The vector $\underline{c}_n$ contains the chip sequence of user n, where $|c_{n,i}|=1$, $i=0,\ldots,N-1$. The elements of the matrix B_n are given by :

$$\left(B_n\right)_{k,\ell} = \frac{1}{N^2} \sum_{i=0}^{N-1} e^{j2\pi \frac{i(k-\ell)}{N}} e^{j\phi_n\left(\frac{iT}{N}\right)} \tag{3}$$

and the additive noise samples $W_{n,0}$ are given by :

$$W_{n,0} = \frac{1}{N} \sum_{k,\ell=0}^{N-1} c_{n,\ell}^* e^{-j2\pi \frac{k\ell}{N}} \int_{-\infty}^{+\infty} n(t) p^*\left(t - \frac{kT}{N}\right) dt \tag{4}$$

The sample at the input of the decision device can be decomposed into four uncorrelated contributions. The first term consists of the mean value, with respect to the carrier phase jitter, of the useful component. The second contribution contains a zero-mean disturbance caused by the fluctuation of the useful component. The third contribution is zero-mean multi-user interference caused by the phase jitter that provokes a loss of orthogonality between the different users. The fourth term is caused by the additive noise.
For small phase jitter $\phi_n(t)$, we can use the approximation $\exp(j\phi_n(t)) \cong 1+j\phi_n(t)$, which reduces the matrix elements $(B_n)_{k,\ell}$ for $k\neq\ell$ to :

$$\left(B_n\right)_{k,\ell} \cong \frac{1}{N^2} \sum_{i=0}^{N-1} j\phi_n\left(\frac{iT}{N}\right) e^{j2\pi \frac{i(k-\ell)}{N}} \tag{5}$$

Similarly, for $k=\ell$

$$\left(B_n\right)_{k,k} \cong \frac{1}{N} + \frac{1}{N^2} \sum_{i=0}^{N-1} j\phi_n\left(\frac{iT}{N}\right) \tag{6}$$

from which we can derive

$$E\left[\left(B_n\right)_{k,k}\right] \cong \frac{1}{N} \tag{7}$$

We define the signal-to-noise ratio (SNR) at the input of the decision device as the ratio of the power of the average useful component to the power of the remaining contributions. In the presence of phase jitter, the SNR is reduced as compared to the case of no jitter $E_{s,n}/E[|W_{n,0}|^2]$. This degradation, expressed in dB, is given by :

$$D_n = 10\log\left(1+\frac{E_{s,n}}{E\left[\left|W_{n,0}\right|^2\right]}E\left[\left|\underline{c}_n^T\left(B_n - E[B_n]\right)\underline{c}_n^*\right|^2\right]+\sum_{\substack{\ell=0\\ \ell\neq n}}^{N-1}\frac{E_{s,\ell}}{E\left[\left|W_{n,0}\right|^2\right]}E\left[\left|\underline{c}_\ell^T\left(B_n - E[B_n]\right)\underline{c}_n^*\right|^2\right]\right) \tag{8}$$

where the power of the fluctuation of the useful component yields :

$$E_{s,n}E\left[\left|\underline{c}_n^T\left(B_n - E[B_n]\right)\underline{c}_n^*\right|^2\right] = E_{s,n}\int_{-\infty}^{+\infty} S_{\phi,n}(f)\left|H_0(f)\right|^2 df \tag{9}$$

and the power of the multi-user interference :

$$\sum_{\substack{\ell=0\\ \ell\neq n}}^{N-1} E_{s,\ell}E\left[\left|\underline{c}_\ell^T\left(B_n - E[B_n]\right)\underline{c}_n^*\right|^2\right] = \frac{1}{N-1}\sum_{\substack{\ell=0\\ \ell\neq n}}^{N-1} E_{s,\ell}\int_{-\infty}^{+\infty} S_{\phi,\ell}(f)\left(1-\left|H_0(f)\right|^2\right)df \tag{10}$$

$S_{\phi,\ell}(f)$ denoting the phase jitter spectrum of user ℓ. The filter $|H_0(f)|^2$ is a low-pass filter :

$$\left|H_0(f)\right|^2 = \left|\frac{1}{N}\frac{\sin(\pi f T)}{\sin(\pi f T/N)}\right|^2 \tag{11}$$

This implies that the fluctuation of the useful component and the multi-user interference mainly contain the low frequency components (<1/T) and the high frequency components (>1/T) of ϕ, respectively. The additive noise yields the power $E[|W_{n,0}|^2]$:

$$E\left[\left|W_{n,0}\right|^2\right] = \int_{-\infty}^{+\infty} S_n(f)\left|P(f)\right|^2 df \tag{12}$$

where P(f) is the Fourier transform of p(t).

In the following we consider the degradation of the SNR under the assumption that all N users have the same jitter spectrum $S_\phi(f)$ and the same energy per symbol E_s. Under these assumptions, the power of the fluctuation of the useful component (9) yields :

$$E\left[\left|\underline{c}_n^T\left(B_n - E[B_n]\right)\underline{c}_n^*\right|^2\right] = E_s\int_{-\infty}^{+\infty} S_\phi(f)\left|H_0(f)\right|^2 df \tag{13}$$

The power of the multi-user interference (10) is given by :

$$\sum_{\substack{\ell=0\\ \ell\neq n}}^{N-1} E_{s,\ell}E\left[\left|\underline{c}_\ell^T\left(B_n - E[B_n]\right)\underline{c}_n^*\right|^2\right] = \frac{M}{N-1}E_s\int_{-\infty}^{+\infty} S_\phi(f)\left(1-\left|H_0(f)\right|^2\right)df \tag{14}$$

where M denotes the number of disturbing users.

Note that for the number of disturbing users equal to N-1, the sum of the powers of the fluctuation of the useful component and multi-user interference only depends on the jitter variance but not on the specific shape of the jitter spectrum : for N-1 disturbing users, the degradation is given by :

$$D_n = 10 \log \left(1 + \frac{E_s}{E\left[|W_{n,0}|^2 \right]} \sigma_\phi^2 \right) \tag{15}$$

where the jitter variance σ_ϕ^2 is defined by :

$$\sigma_\phi^2 = \int_{-\infty}^{+\infty} S_\phi(f) df \tag{16}$$

COMPUTATIONAL RESULTS

In the simulations, we made use of the orthogonal Walsh-Hadamard chip sequences. We subjected the system to phase jitter $\phi(t)$, having a power spectral density $S_\phi(f)$ shown in Fig. 2. The psd is dimensioned in such a way that the jitter variance equals σ_ϕ^2. Fig. 3 shows the degradation as function of the normalized number of disturbing users M/(N-1) for $f_L=0$, $f_H=B$, N=32, $\sigma_\phi^2=10^{-4}$ and the $SNR_{\phi=0}$, which corresponds to $E_s/E[|W_{n,0}|^2]$, equals 25dB. Enlarging the jitter bandwidth B reduces the power of the fluctuation of the useful component and increases at the same time the power of the multi-user interference, which engenders a stronger dependency on the number of disturbing users. For N-1 disturbing users, the degradation becomes only dependent on the jitter variance and not on the shape of the jitter spectrum.

The scatter diagrams for OFDM/CDMA-QPSK, shown in Figs. 4 and 5, are made in absence of additive noise, for $f_L=0$, $f_H=2.56/T$ and $f_L=2.56/T$, $f_H=5.12/T$ respectively. In both cases, the jitter variance was taken σ_ϕ^2=-21dB and the number of carriers N=128. Figs. 4 and 5 demonstrate that the spectral distribution of the jitter determines the outlook of the scatter diagram. This can be explained as follows : the fluctuation of the useful component, which is mainly determined by the low frequency components (<1/T) of the phase jitter, gives rise to an essentially angular displacement of the sample at the input device, while the radial displacement is negligible. As easily can be proved, the term of the multi-user interference has uncorrelated real and imaginary part, which implies that this contribution shows a circular distribution. In Fig. 4, the jitter spectrum mainly contains low-frequency components, which signifies that the phase jitter essentially provokes a fluctuation of the useful component, whereas in Fig. 5, no low-frequency components are present, so that the multi-user interference becomes the major contribution. This considerations explain easily the nature of the scatter diagrams.

CONCLUSIONS

We have investigated the influence of carrier phase jitter on the performance of an OFDM/CDMA system. The results can be summarized as follows :

- The degradation, caused by the phase jitter, depends on the number of disturbing users. For a given jitter variance, the degradation is for phase jitter with low spectral contents less dependent of the number of disturbing users than for phase jitter without low frequency components.
- For N-1 disturbing users, the degradation only depends on the jitter variance, not of the spectral contents of the jitter.
- The jitter components at low frequencies (<1/T) cause essentially an angular displacement of the sample at the input of the decision device, while the jitter components at higher frequencies give rise to an angular as well as a radial displacement.

REFERENCES

[1] L.M.A. Jalloul, J.M. Holtzman, "Performance Analysis of DS/CDMA with. Noncoherent M-ary Orthogonal Modulation in Multipath Fading Channels", IEEE J. on Sel. Areas in Comm., vol. 12, no 5, Jun 94, pp. 862-870

[2] E.A. Sourour, M. Nakagawa, "Performance of Orthogonal Multicarrier CDMA in a Multipath Fading Channel", IEEE Trans. On Comm., vol. 44, no 3, Mar 96, pp. 356-367

[3] V.M. Da Silva, E.S. Sousa, "Multicarrier Orthogonal CDMA Signals for Quasi-Synchronous Communication Systems", IEEE J. on Sel. Areas in Comm., vol. 12, no 5, Jun 94, pp. 842-852

[4] S. Hara, T.H. Lee, R. Prasad, "BER comparison of DS-CDMA and MC-CDMA for Frequency Selective Fading Channels", Proc. 7th Thyrrenian Workshop on Digital Communications, Viareggio Italy Sep 95, Springer, pp. 3-14

[5] L. Vandendorpe, O. van de Wiel, "Decision Feedback Multi-User Detection for Multitone CDMA Systems", Proc. 7th Thyrrenian Workshop on Digital Communications, Viareggio Italy Sep 95, Springer, pp. 39-52

[6] Y. Sanada, M. Nakagawa, "A Multiuser Interference Cancellation Technique Utilizing Convolutional Codes and Orthogonal Multicarrier Communications", IEEE J. on Sel. Areas in Comm., vol. 14, no 8, Oct 96, pp. 1500-1509

[7] T. Pollet, M. Moeneclaey, I. Jeanclaude, H. Sari, "Comparison of Single-Carrier and Multi-Carrier QAM System Performance in the Presence of Carrier Phase Jitter", WIRE139-SI

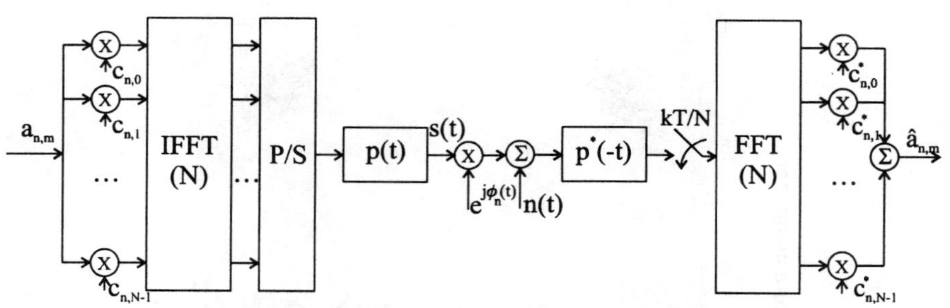

Fig. 1 : Conceptual block diagram of an OFDM/CDMA transceiver

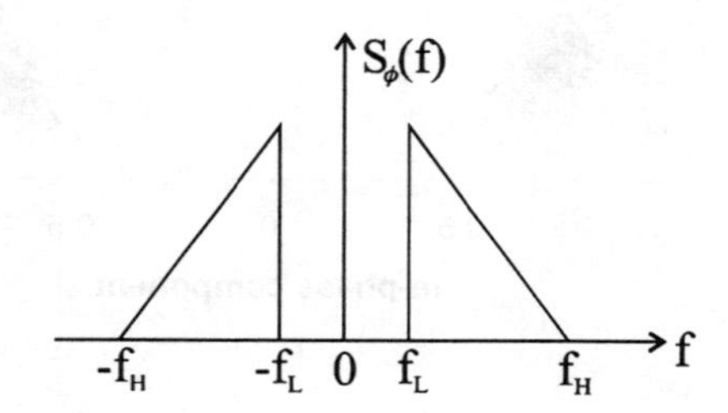

Fig. 2 : Phase jitter spectrum

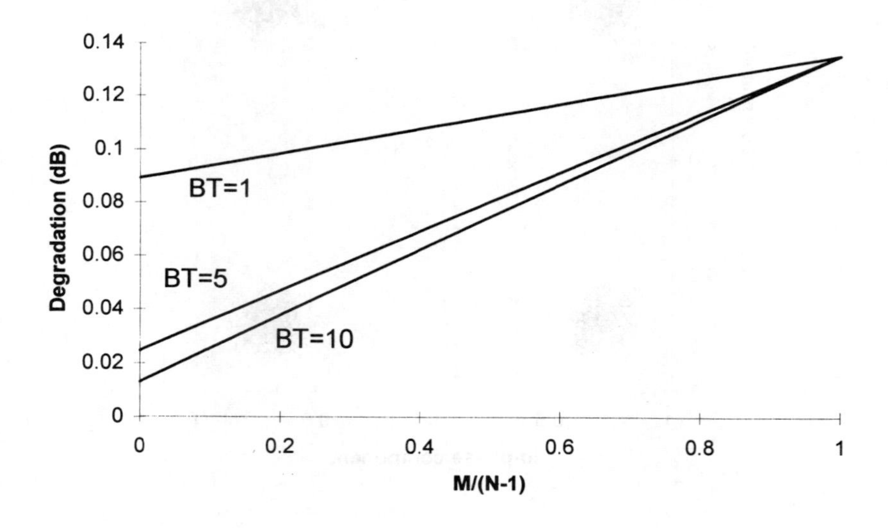

Fig. 3 : Degradation as function of M : $s_\phi^2=10^{-4}$, $SNR_{\phi=0}$=25dB

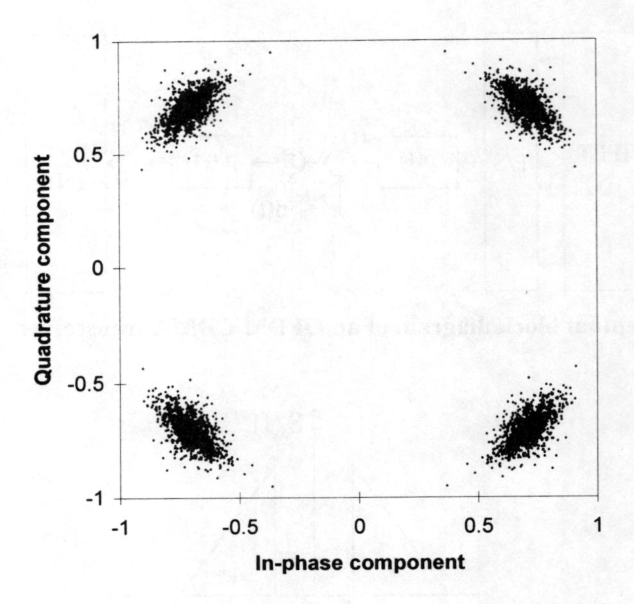

Fig. 4 : Scatter diagram for f_L=0, f_H=2.56/T, N-1 disturbing users

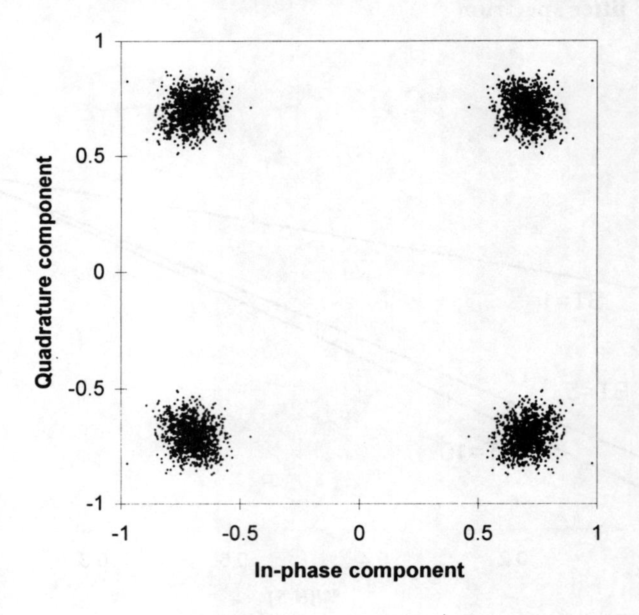

Fig. 5 : Scatter diagram for f_L=2.56/T, f_H=5.12/T, N-1 disturbing users

Time Domain Uplink Channel Estimation in Multicarrier–CDMA Mobile Radio System Concepts

Bernd Steiner
Deutsche Telekom AG, Technologiezentrum Darmstadt,
P.O. Box 10 00 03, D–64276 Darmstadt, Germany
Phone: +49 6151 83 2519, Fax: +49 6151 83 4325,
E-mail: steinerb@tzd.telekom.de

Abstract

The combination of OFDM (Orthogonal Frequency Division Multiplexing) and CDMA (Code Division Multiple Access) has recently been proposed for mobile radio transmission. The present paper considers the issue of MIMO (Multiple Input Multiple Output) channel estimation in the uplink of an OFDM–CDMA transmission system. The channel estimation is based on pilot tones which are transmitted with the subcarriers used for data transmission. A method of constructing the pilot tones of the simultaneously active users from one basic sequence is proposed. This method facilitates both a high quality of the estimated channel transfer functions and a low–complexity channel estimator. Simulation results including a realistic channel model show that the proposed channel estimation technique is well–suited to attain a high spectral capacity.

1 INTRODUCTION

The application of multicarrier (MC) data transmission techniques was first proposed by Weinstein and Ebert in [1]. When the paper was published, a practical application of MC techniques, however, was not thought of since MC necessitates DFT and IDFT performed in real time. With the tremendous development in microelectronics, state of the art DSPs (Digital Signal Processors) are capable of performing this operation. Consequently, a multicarrier modulation, i.e. OFDM (Orthogonal Frequency Division Multiplexing), was chosen for the terrestrial transmission in DAB (Digital Audio Broadcasting) [2] and DVB (Digital Video Broadcasting) [3], respectively.

The combination of MC techniques and CDMA was first addressed in [4]. Since then, numerous publications dealt with the various aspects of MC–CDMA transmission. CDMA is a multiple access scheme and is therefore applicable in multiple access systems where a number of users bidirectionally communicate with a base station. Some examples for multiple access systems using bidirectional radio communications are mobile radio systems, wireless local loop (WLL) systems and wireless local area networks (WLANs). Most publications dealing with MC–CDMA communication systems focus on the downlink. It is, however, tempting to use both the same modulation scheme and the same multiple access scheme in the uplink and the downlink of the considered communications system. In the present paper, the more elaborate uplink of an OFDM–CDMA (being a special case of MC–CDMA) mobile radio system concept is considered. The main focus is the issue of high resolution multiple input multiple output (MIMO) channel estimation which must precede a coherent data detection.

The present paper is organized as follows: Sect. 2 gives a description of the system model considered in the paper. Sect. 3 addresses the problem of channel estimation. Here, the

K. Fazel and G.P. Fettweis (eds.), Multi-Carrier Spread-Spectrum, 153-160.

presuppositions which are necessary to facilitate a MIMO channel estimation in an OFDM-CDMA system are discussed. Furthermore, the general approach to attain a high resolution channel estimate by least squares estimation is sketched. Additionally, a method of specially selecting the different users' pilot tones is proposed. This facilitates a low cost channel estimation. Sect. 4 gives performance results for a realistic mobile radio channel model. Finally, a conclusion of the present paper is given in Sect. 5.

2 SYSTEM MODEL

In the present paper, the uplink of an OFDM–CDMA mobile radio system concept employing burst transmission in considered. Each burst consists of a single OFDM–CDMA symbol. As usual in OFDM transmission, an appropriately chose guard interval avoids ISI (Inter Symbol Interference) and ICI (Inter Carrier Interference). Since the transmission occurs in bursts, a TDMA component may easily be introduced adding a further degree of flexibility.

The number of subcarriers in each OFDM–CDMA symbol is P. Since a discrete–frequency formalism is applied, it is sufficient to characterize a specific subcarrier by its index. The index of the subcarrier with the lowest center frequency is 1 and the index of the one with the largest center frequency is P. Within each OFDM-CDMA symbol, N data symbols are transmitted. Due to the spectral spreading, each data symbol is transmitted via Q subcarriers. Therefore, NQ subcarriers are used for the data transmission. The issues of data modulation and spectral spreading shall not be dealt with in the present paper, for details see [6]. In the following, it is assumed that $P < NQ$ holds so that a number of subcarriers can be used for the channel estimation. With the pilot tone spacing N_{CE}, the subcarriers whose index fulfill $\mathrm{frac}((i-\mu)/N_{\mathrm{CE}}) = 0$ are used as pilot tones, i.e. pilot tones are inserted regularly to facilitate a channel estimation ($\mathrm{frac}(\cdot)$ denotes the fractional part of a real number). Fig. 1 depicts the transmission of pilot tones with the subcarriers used for data transmission for N_{CE}=4 and P=20. For a practical implementation, the total number P of subcarriers is considerably larger than 20 and preferably a power of 2. Without loss of generality, it is assumed that P/N_{CE} is integer and that P/N_{CE} subcarriers are used for the channel estimation. Since the center frequencies of the pilot tones are given by the pilot tone spacing N_{CE} and the parameter μ, the only degree of freedom is the complex amplitude of a considered pilot tone. Consequently, the pilot tones are determined by μ, N_{CE} and P/N_{CE} complex amplitudes denoted by $m_i^{(k)}$. Now the discrete–frequency sequence

$$t_{\mathrm{CE},i}^{(k)} = \begin{cases} m_{(1+(i-\mu)/N_{\mathrm{CE}})}^{(k)} & \mathrm{frac}((i-\mu)/N_{\mathrm{CE}}) = 0, \\ 0 & \text{else}, \end{cases} \tag{1}$$

whose nonzero elements are associated with the pilot tones, is introduced. The entire discrete–frequency sequence of the kth user is termed $t_i^{(k)}$. Furthermore, the discrete–frequency sequence associated with the subcarriers used for data transmission is termed $t_{\mathrm{D},i}^{(k)}$. Obviously, $t_{\mathrm{CE},i}^{(k)} t_{\mathrm{D},i}^{(k)} = 0$, $i = 1 \ldots P$, holds. In the present paper, it is assumed that all elements of $t_i^{(k)}$ have the same magnitude. Therefore, the only degree of freedom of a specific pilot tone is its phase $\arg(t_{\mathrm{CE},i}^{(k)})$. As usual in OFDM transmission system, an IDFT (preferably performed a IFFT) of the sequence $t_i^{(k)}$ is formed. After supplying the guard interval, the resulting discrete–time signal is D/A converted at a rate of R=B with B being the system bandwidth.

At the receiver, the received signal is band limited, down converted and sampled at a rate of R=B. Now, the DFT (usually performed as an FFT) of P appropriately chosen samples is considered. Since both ISI and ICI are not present, the P values of the above mentioned sequence are determined by

$$r_i = n_i + \sum_{k=1}^{K} t_i^{(k)} h_i^{(k)}, i = 1 \ldots P. \tag{2}$$

In eq. (2), n_i is a noise sequence and $h_i^{(k)}$ is associated with the samples of the transfer function of the radio channel between the kth user and the base station. Note, that eq. (2)

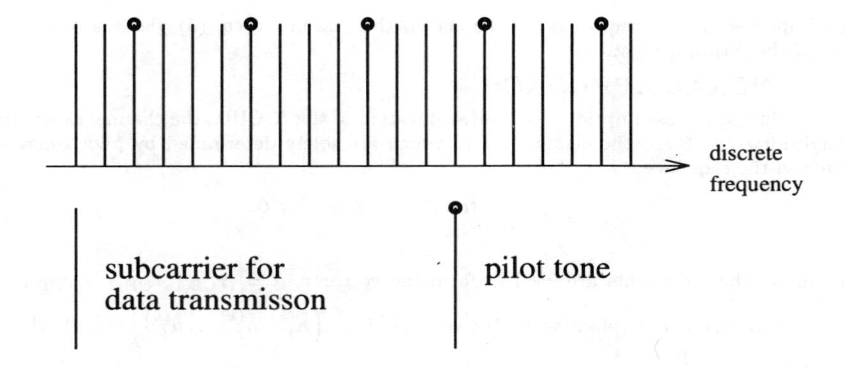

Fig. 1: Transmission of pilot tones with the subcarriers for data transmission for N_{CE}=4 and P=20

is only valid when the contributions of *all* users to received signal do not lead to ISI or ICI, respectively. This constraint necessitates a mutual synchronization of the transmit signals of the synchronously active users. Therefore, the proposed concept is a synchronous OFDM–CDMA mobile radio system concept comparable with the JD–CDMA mobile radio system concept [8, 9].

3 CHANNEL ESTIMATION

3.1 PRESUPPOSITIONS

The issue of channel estimation in a radio transmission system is a prerequisite for a coherent data detection. When the time–variance of the radio channels is small with respect to the duration of an OFDM symbol, an iterative channel estimation techniques, e.g. proposed in [10], can be used. Due to the considerable time–variance which may be present in mobile radio systems, an iterative approach shall not be dealt with in the present paper. Rather than that, a "singe shot" channel estimation technique is necessary, i.e. the K channel transfer functions shall solely be estimated from the sequence r_i according to eq. (2). Therefore, the P/N_{CE} elements of r_i which are associated with the nonzero elements of $t_{\mathrm{CE},i}^{(k)}$ may be exploited for channel estimation. However, a total of KP samples of the transfer functions must be determined. When least squares channel estimation techniques shall be applied, this implies the determination of KP unknown samples of the transfer functions from P/N_{CE} observations by solving a system of equations. Obviously, this aim cannot be reached when no further knowledge concerning the transfer functions to be estimated is available.

However, the problem of channel estimation can be solved, when a time domain channel estimation approach is used. As mentioned before, a guard interval is utilized to avoid ISI and ICI. In the following it is supposed that the maximum duration of a channel impulse response (CIR) is bounded by the duration of the guard interval. Due to the limited duration of a CIR, adjacent samples of a transfer function, e.g. $h_i^{(k)}$ and $h_{i+1}^{(k)}$ are not independent. Therefore, the number of unknowns can be reduced to facilitate an unbiased channel estimation via least squares estimation. In what follows, it is supposed that

$$\left(h_1^{(k)}, h_2^{(k)} \ldots h_P^{(k)}\right) = \mathrm{DFT}\left(\tilde{h}_1^{(k)}, \tilde{h}_2^{(k)} \ldots \tilde{h}_W^{(k)}, 0, 0 \ldots 0\right) \tag{3}$$

holds for the samples of the channel transfer function of a specific user. The tilde denotes the elements of sequences in the discrete–time domain. Therefore, the number of unknown is reduced to KW. Note that eq. (3) is normally not valid since a CIR is not a train of equally spaced impulses. In the sequel, it will be shown that assuming eq. (3) yields a reasonable quality of the estimates, yet.

3.2 GENERAL APPROACH

After introducing the assumption of a limited duration of the K CIRs, the channel estimation is straightforward. First, the elements of r_i which are solely determined by pilot tones are arranged in the sequence

$$r_{\mathrm{CE},i} = \begin{cases} r_i & \mathrm{frac}((i-\mu)/N_{\mathrm{CE}}) = 0, \\ 0 & \text{else.} \end{cases} \tag{4}$$

Furthermore, these elements are used to form the vector $\boldsymbol{r}_{\mathrm{CE}} = \left(r_{\mathrm{CE},1}, r_{\mathrm{CE},2} \ldots r_{\mathrm{CE},P}\right)^{\mathrm{T}}$. In order to describe the relationship between $\tilde{\boldsymbol{h}}^{(k)} = \left(\tilde{h}_1^{(k)}, \tilde{h}_2^{(k)} \ldots \tilde{h}_P^{(k)}\right)^{\mathrm{T}}$ and $\boldsymbol{h}^{(k)} = \left(h_1^{(k)}, h_2^{(k)} \ldots h_P^{(k)}\right)^{\mathrm{T}}$, the unitary $P \times P$ matrix $\boldsymbol{D}$ of the discrete Fourier transform is introduced. Now, eq. (3) may equivalently be expressed by $\boldsymbol{h}^{(k)} = \boldsymbol{D}\tilde{\boldsymbol{h}}^{(k)}$. Since only W elements of $\tilde{\boldsymbol{h}}^{(k)}$ are nonzero, these elements are used to form the vector $\tilde{\boldsymbol{h}}_W^{(k)} = \left(\tilde{h}_1^{(k)}, \tilde{h}_2^{(k)} \ldots \tilde{h}_W^{(k)}\right)^{\mathrm{T}}$. The sub-matrix that comprises the first W columns of $\boldsymbol{D}$ shall be labeled $\boldsymbol{D}_W$. Hence, $\boldsymbol{h}^{(k)} = \boldsymbol{D}_W \tilde{\boldsymbol{h}}_W^{(k)}$ holds. To take into account the effect of the complex amplitude of the pilot tones on the received signal, the diagonal matrix

$$\boldsymbol{G}^{(k)} = \mathrm{Diag}\{\boldsymbol{G}^{(k)}\},\ G_{ii}^{(k)} = t_{\mathrm{CE},i}^{(k)} \tag{5}$$

is introduced. Note that $\boldsymbol{G}^{(k)}$ has only P/N_{CE} nonzero entries. With the noise vector $\boldsymbol{n}$ which also only has P/N_{CE} nonzero elements

$$\boldsymbol{r}_{CE} = \boldsymbol{n} + \sum_{k=1}^{K} \boldsymbol{G}^{(k)} \boldsymbol{D}_W \tilde{\boldsymbol{h}}^{(k)} \tag{6}$$

holds.

A single-user estimate of $\boldsymbol{h}^{(k)}$ may be determined by matched filtering, i.e.

$$\hat{\boldsymbol{h}}^{(k)} = \boldsymbol{D}\boldsymbol{D}_W^{\mathrm{H}} \boldsymbol{G}^{(k)\mathrm{H}} \boldsymbol{r}_{CE}. \tag{7}$$

Using a matrix vector notation, the multiplication by the diagonal elements of $\boldsymbol{G}^{(k)\mathrm{H}}$ provides matched filtering and therefore removes the effect of the phase of the pilot tones. The multiplication by the matrix $\boldsymbol{D}\boldsymbol{D}_W^{\mathrm{H}}$ provides a spectral smoothing which is necessary to attain estimates for the samples of a transfer function where the subcarriers for data transmission are present. The estimate according to eq. (7) is corrupted by multiple access interference (MAI) in the multiuser case. To avoid MAI, the $P \times KW$ matrix

$$\boldsymbol{V} = (\boldsymbol{G}^{(1)} \boldsymbol{D}_W, \boldsymbol{G}^{(2)} \boldsymbol{D}_W \ldots \boldsymbol{G}^{(K)} \boldsymbol{D}_W) \tag{8}$$

is introduced. Starting from $\boldsymbol{V}$, the well–known estimation matrix [11]

$$\boldsymbol{M} = (\boldsymbol{V}^{\mathrm{H}} \boldsymbol{V})^{-1} \boldsymbol{V}^{\mathrm{H}} \tag{9}$$

used for least squares estimation is computed. The matrix $\boldsymbol{V}^{\mathrm{H}} \boldsymbol{V}$ must be positive definite to assure the existence of the estimation matrix $\boldsymbol{M}$. To attain estimates not perturbed by MAI, first the time domain estimate $\hat{\tilde{\boldsymbol{h}}} = \boldsymbol{M}\boldsymbol{r}_{CE}$ is determined. Then, the desired estimates of the samples of the channel transfer function are given by applying a DFT to the time domain estimate according to

$$\hat{\boldsymbol{h}}^{(k)} = \boldsymbol{D}_W (\hat{\tilde{h}}_{1+W(k-1)}, \hat{\tilde{h}}_{2+W(k-1)} \ldots \hat{\tilde{h}}_{kW})^{\mathrm{T}},\ k = 1 \ldots K. \tag{10}$$

Fig. 2 depicts the structure of the channel estimator with respect to a practical implementation, i.e. the multiplication with $\boldsymbol{D}_W^{\mathrm{H}}$ is implemented as an FFT. It is obvious from Fig. 2 that the channel estimation is a rather complex operation since it necessitates K IFFTs, the multiplication of a full $KW \times KW$ matrix by a vector having KW elements and K FFTs to compute the desired transfer functions from the time domain estimates. As indicated in Fig. 2, the multiplication of $\hat{\tilde{\boldsymbol{h}}}$ by $(\boldsymbol{V}^{\mathrm{H}}\boldsymbol{V})^{-1}$ provides a time domain MAI elimination. In the single–user case, however, the matrices $(\boldsymbol{V}^{\mathrm{H}}\boldsymbol{V})^{-1}$ and $\boldsymbol{V}^{\mathrm{H}}\boldsymbol{V}$ are diagonal matrices due to the orthogonality of the pilot tones. Therefore, MAI elimination is not necessary in the single user case. This will be exploited in the next subsection.

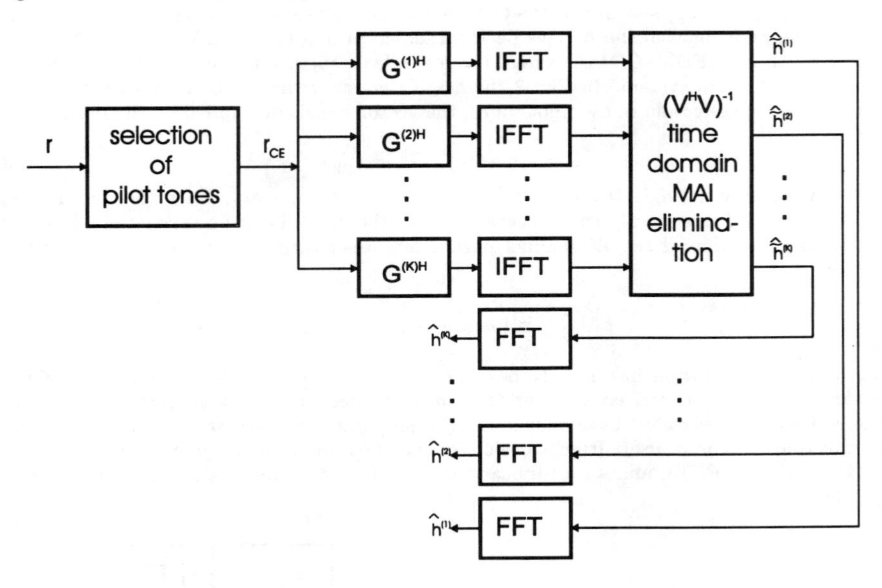

Fig. 2: General structure of the channel estimator for synchronous OFDM–CDMA transmission systems

Another drawback of the channel estimator besides its complexity is the SNR degradation of the time domain estimates which occurs due to the noise enhancement of the least squares estimation. The SNR degradation associated with the uth element of $\hat{\tilde{\boldsymbol{h}}}$ is determined by [7, 8]

$$d_u/\mathrm{dB} = 10\log_{10}\left([(\boldsymbol{V}^{\mathrm{H}}\boldsymbol{V})^{-1}]_{u,u}[(\boldsymbol{V}^{\mathrm{H}}\boldsymbol{V})]_{u,u}\right). \tag{11}$$

When the average SNR degradation is e.g. 3 dB, 50% of the energy used for the channel estimation is lost. With respect to the impact of the non perfect channel estimation on the detected data it is desirable to maximize the energy efficiency of the channel estimator and hence to minimize the SNR degradation. When a high system capacity is aimed at, however, $KW \approx P/N_{\mathrm{CE}}$ holds. In this case, the SNR degradations are in the range of a few dB.

3.3 SPECIAL SELECTION OF THE PILOT TONES

Both the above mentioned SNR degradations and the complexity of the channel estimator can be reduced, when the channel estimation technique proposed for the JD–CDMA mobile radio system concept is used [7]. The basic idea of this channel estimation technique is to derive the training sequences of the synchronously active users from one basic sequence and to reduce the problem of MIMO channel estimation to channel estimation in the single user case. In the proposed OFMD–CDMA transmission concept this aim can be reached when the pilot tones of the users $2 \ldots K$ are derived from the pilot tones of user 1 according to [6]

$$t_{\mathrm{CE},i}^{(k)} = t_{\mathrm{CE},i}^{(1)} \exp\left(-j2\pi W \frac{(k-1)(i-1)}{P}\right). \tag{12}$$

From an illustrative point of view eq. (12) implies that, if only the pilot tones for channel estimation were transmitted, the transmit signal of the kth user is delayed version of the first user's transmit signal [6]. The delay is $(k-1)W/B$. Note, that this is the same channel estimation principle applied in the JD–CDMA mobile radio system concept [7]. Due to the fact that the pilot tones of the K have been chosen according to eq. (12), the MIMO channel estimation for an OFDM–CDMA system has a similar structure as the channel estimator in a single–user OFDM system. In Fig. 3 the associated structure of the channel estimator is shown. After the selection of the pilot tones, the vector $\boldsymbol{r}_{\mathrm{CE}}$ is multiplied by the matrix $\boldsymbol{G}^{\mathrm{H}}$, with

$$\boldsymbol{G} = \mathrm{Diag}\{\boldsymbol{G}\},\ G_{ii} = t_{\mathrm{CE},i}^{(1)}. \tag{13}$$

Furthermore, only a single IFFT instead of K IFFTs is necessary, since the first P/N_{CE} values of the FFT are simply the concatenation of the K CIRs to be estimated. The next step is the extraction of the W channel parameters associated with the discrete–time CIR according to

$$\left(\hat{\hat{h}}_1^{(k)}, \hat{\hat{h}}_2^{(k)} \ldots \hat{\hat{h}}_W^{(k)}\right)^{\mathrm{T}} = \left(\hat{\hat{h}}_{(k-1)W+1}, \hat{\hat{h}}_{(k-1)W+2} \cdots \hat{\hat{h}}_{kW}\right)^{\mathrm{T}}. \tag{14}$$

Note, that this operation has already been implied in eq. (10). After extracting the CIRs, the samples of the K desired transfer functions are determined by performing K FFTs. Irrespective of the selection of the pilot tones, this step is always necessary. However, deriving the pilot tones of the K users from a basic sequence facilitates the channel estimation, since $K-1$ additional IFFTs, and a multiplication of a vector by a full matrix are not necessary any more.

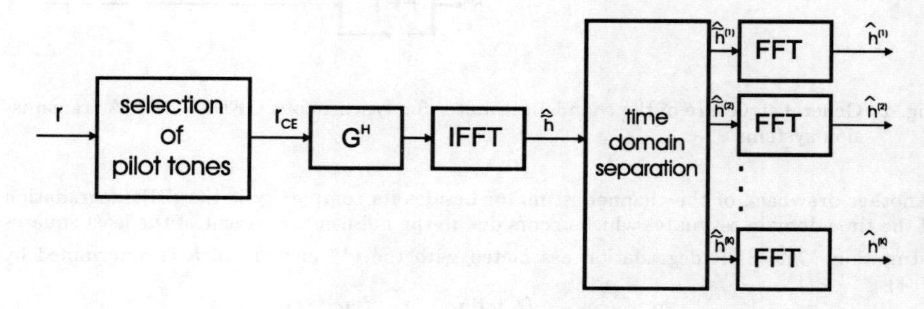

Fig. 3: Structure of the channel estimator when the pilot tones are derived from a basic sequence

The derivation of the pilot tones of the different users from a single basic sequence also avoids the SNR degradations associated with the time–domain estimates, cf. Sect. 3.2. It was already mentioned before that MAI does not occur in the single users case. Consequently, no MAI elimination leading to a noise enhancement and therefore to an SNR–degradation is necessary. Hence, the energy efficiency of the proposed channel estimator is considerably better than in the case of independently choosing the pilot tones.

4 PERFORMANCE OF THE CHANNEL ESTIMATOR

4.1 SIMULATION PARAMETERS

To evaluate the performance of the channel estimation method proposed in the present paper, simulations were made. The following simulation parameters have been assumed:

$$\begin{array}{c} P = 1024,\ Q = 16,\ N = 47, \\ N_{\mathrm{CE}} = 4,\ K = 8, W = 25. \end{array} \tag{15}$$

QPSK is used for data modulation. With these values of P and N_{CE}, a maximum of 256 channel parameters can be estimated in a single OFDM–CDMA symbol. The power delay profile of the WSSUS channels is supposed to be

$$\rho(\tau) = \exp(-0,48B\tau)\mathrm{rect}\left(\frac{\tau}{25B}\right). \tag{16}$$

When the system bandwidth B equals 2MHz, the channel model complies well with the ***typical urban*** channel model according to COST 207 [12]. The transmission channels were time–invariant within the duration of an OFDM–CDMA symbol.

4.2 SIMULATION RESULTS

In Sect. 3.1 is was stated that the channel estimation becomes possible when a discrete–time CIR with W nonzero channel parameters is supposed. A real–world CIR is not a discrete–time function and can therefore not be exactly be decribed by the channel model. This gives rise to systematic channel estimation errors which cannot be dealt with in detail in the present paper due to space considerations. In [6], however, the effect of the systematic errors is depicted for a sample channel transfer function. Besides these systematic errors, an estimated channel transfer function is also corrupted by additive noise. As already mentioned, simulations were made to evaluate the impact of both the systematic errors and the additive noise, cf. Fig. 4, which shows the uncoded bit error rate BER versus the average SNR per bit. The well–known detection technique MMSE–BDFE [6, 8] was used for multiuser detection. Besides the case of a real channel estimation, the cases of perfect channel estimation, of a channel estimation not corrupted by noise but by systematic errors as well as a channel estimation corrupted by noise but not by systematic errors were considered. Systematic channel estimation errors may lead to an irreducible error floor. It is obvious from Fig. 4 that the irreducible bit error rate is below 10^{-3}. A typical uncoded target bit error rate necessary in mobile radio transmission is 1%. It follows from Fig. 4 that this target bit error rate is met at an SNR of approximately 10.5 dB in the case of a real channel estimation. When no systematic channel estimation errors are present, the necessary SNR is reduced to 9.9 dB. For a perfect channel estimation, the SNR is 7 dB. Therefore, the additive noise rather than the systematic channel estimation errors is the main source of the SNR degradation due to the non perfect channel estimation. The SNR degradation due to the channel estimation is comparable with the one that occurs in a JD–CDMA system concept employing DS–CDMA [7] with a small processing gain.

5 CONCLUSION

In the present paper, the issue of a pilot tone based uplink channel estimation in a synchronous OFDM–CDMA mobile radio system concept is considered. The problem of simultaneously estimating several channel transfer functions is solved by introducing a time domain channel estimation. The structure of the channel estimator is rather complex and its performance with respect to the noise enhancement is rather poor when the pilot tones of the simultaneously active users are selected independently. By deriving these pilot tones from a single sequence, the complexity of the channel estimator is reduced considerably and the performance with respect to the noise enhancement is optimum. Simulations performed for a realistic channel model show that the performance of an OFDM-CDMA system employing the proposed channel estimator is comparable with the performance of a JD–CDMA system concept using the same system parameters. Therefore, OFDM–CDMA is a well–suited candidate for the air interface of next generation mobile radio systems.

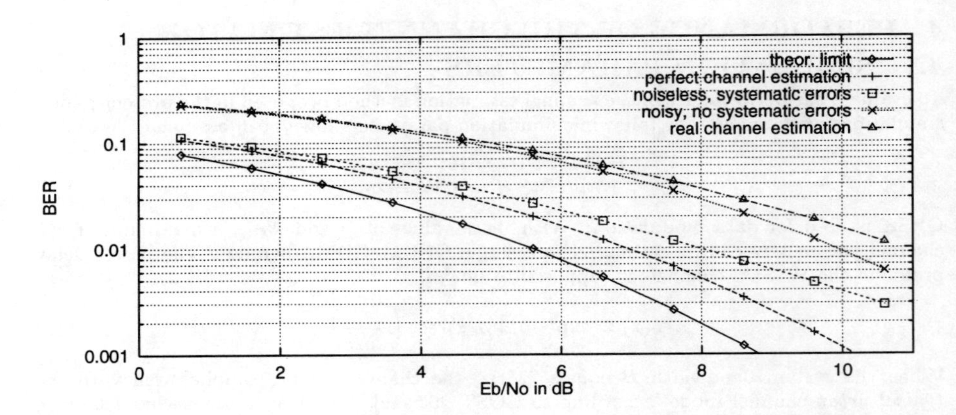

Fig. 4: Bit error rate BER versus the average SNR per bit of an OFDM–CDMA mobile radio concept with 8 synchronously active users with different degrees of non perfect channel estimation

References

[1] Weinstein, S.: Ebert, P.M.: Data transmission by frequency–division multiplexing using the discrete fourier transform. IEEE Transactions on Communications, vol. 40, pp. 628–634, 1971.

[2] O'Leary, T.: Terrestrial digital audio broadcast in Europe. EBU Technical Review, pp. 19–26, 1993.

[3] Monnier, R.; Rault, J.B.; de Couasnon, T.: Digital television broadcasting with high spectral efficiency. Report of IBC, Amsterdam, pp. 380-384, 1992.

[4] Yee, N.; Linnartz, J.–P.; Fettweis, G.: Multi–carrier CDMA in indoor wireless radio networks. Proc. International Symposium on Personal, Indoor and Mobile Radio Communications, (PIMRC '93), Yokohama, pp. 109-113, 1993.

[5] Fazel, K.; Kaiser, S.; Schnell, M.: A flexible and high performance cellular communications system based on orthogonal multi–carrier SSMA. International Journal on Wireless Personal Communications, pp. 121–144, 1995.

[6] Steiner, B.: Uplink performance of a multicarrier–CDMA mobile radio system concept. To appear in Proc. Vehicular Technology Conference (VTC '97), Phoenix, 1997.

[7] Steiner, B.: Ein Beitrag zur Mobilfunk–Kanalschätzung unter besonderer Berücksichtigung synchroner CDMA–Mobilfunksysteme mit Joint Detection (in German). Ph.D. Thesis, VDI Fortschritt–Berichte, Reihe 10, Nr. 337, Düsseldorf, 1995.

[8] Klein, A.: Multi–user detection of CDMA signals — algorithms and their application to cellular mobile radio. Ph.D. Thesis, VDI Fortschritt–Berichte, Reihe 10, Nr. 423, Düsseldorf, 1996.

[9] Jung, P.; Steiner, B.: A joint detection CDMA mobile radio system concept developed within COST 231. Proc. Vehicular Technology Conference (VTC '95), pp. 469–473, 1995.

[10] Cacopardi, S.; Fedele, G.; Frescure, F.; Realti, G.: A frequency division duplex configuration for indoor wireless multicarrier DS–CDMA systems. Proc. EPMCC '95, Bologna, pp. 237–244, 1995.

[11] Whalen, A.D.: Detection of signals in noise. Academic Press, San Diego, 1971.

[12] COST 207: Digital land mobile radio communications. Final Report. Office for Official Publications of the European Communities, Luxemburg, 1989.

SUBSPACE-BASED JOINT TIME-DELAY AND FREQUENCY-SHIFT ESTIMATION IN MULTITONE-CODE DIVISION MULTIPLE ACCESS (MT-CDMA) SYSTEMS

Miljko Erić[1], Milorad Obradović[2], Igor Simić[3]

[1]Institute of Electrical Engineering, VTI, Belgrade, Katanićeva 15, Yugoslavia
E-mail: meric@galeb.etf.bg.ac.yu

[2]Faculty of Technical Science, Novi Sad, Yugoslavia

[3]IMTEL - Institute of Microwave Techniques and Electronics, B. Lenjina 165b, 11 070 Belgrade, Yugoslavia

ABSTRACT:

In this paper we proposed a subspace-based algorithm for joint time-delay and frequency shift estimation in multitone-code division multiple access (MT-CDMA) systems. Proposed algorithm is near-far robust, requires no preamble and is well suited to fading channels.

K. Fazel and G.P. Fettweis (eds.), Multi-Carrier Spread-Spectrum, 161-167.

1. INTRODUCTION

Near-far problem and maintaining orthogonality conditions are serious problems in asynchronous CDMA systems. Recently, there has been increased interest in subspace-based methods for near-far time delay estimation and synchronization in asynchronous CDMA systems [1,2,3,4], and channel estimation [5,6]. Subspace-based methods are near-far robust (resistant) and can be applied when orthogonality condition is not perfectly satisfied. As we know, up to now subspace-based methods are not applied in Multi-Carrier CDMA systems. In this paper we considered a subspace-based joint time-delay and frequency shift estimation in MT CDMA systems. Novelty in the proposed subspace-based concept is in fact that signal vectors depends on unknown time-delay and frequency shift which is due to subcarrier oscillator errors or Doppler shift in mobile environment and known frequency subcarrier separation. Proposed estimator is near-far robust, requires no preamble and is well suited to fading channels.

2. SYSTEM MODEL

System under consideration is Multi-Tone Code Division Multiple Access (MT CDMA) system [7,8] with K active users and N_C subcarriers per user. Baseband signal of the k*th* user and i*th* subcarrier $s_k^{(i)}(t)$ can be defined as:

$$s_k^{(i)}(t) = \sum_{m=-\infty}^{\infty} d_k^{(i)}(m) b_k(t - mT_s). \tag{1}$$

In eq.1. $T_S = N_C T$ is bit duration on the subcarriers (after serial to parallel conversion); Nc is number of subcarriers; T is bit duration of input data stream (before serial to parallel conversion). Subscript k and superscript i imply that subscripted and superscripted quantity is due to k*th* user and i*th* subcarrier respectively; $b_k(t)$ is a k*th* user code waveform, where $b_k(t)=0$ for $[0,T_S)$; T_C is chip duration where $T_C=T_S/N$; N is integer (length of spreading code waveform). Code waveforms $b_k(t)$ are assumed to be rectangular, unit amplitude and periodic with period T_S; $d_k^{(i)}$ is data stream on the i*th* subcarrier.

Transmitted signal of the k*th* user is formed as:

$$s_k(t) = \sum_{i=0}^{N_c-1} s_k^{(i)}(t)\sqrt{2P_k^{(i)}} \cos(\omega_0 t + i\Delta\omega_{sc} t + \theta_k^{(i)})) \tag{2}$$

where $P_k^{(i)}$ is power of the i*th* subcarrier of k*th* user, ω_0 is radial carrier frequency, $\Delta\omega_{sc}$ is frequency subcarrier separation and $\theta_k^{(i)}$ is random phase of i*th* subcarrier.

Received signal which is superposition of K active users can be written as:

$$r(t)=\mathrm{Re}\{\sum_{k=1}^{K}\sum_{i=0}^{N_c-1}s_k^{(i)}(t-\tau_k^{(i)})\sqrt{2P_k^{(i)}}\exp(j(\omega_0 t+i\Delta\omega_{sc}t+\Delta\omega_{Dk}^{(i)}t+\theta_k^{(i)''}))\}+n(t)$$

(3)

where: $n(t)$ is additive white Gaussian noise width two-sided power spectral density $N_0/2$; $\Delta\omega_{Dk}^{(i)}$ is frequency shift of the ith subcarrier which is due to oscillators errors or Doppler shift in mobile environment.

Suppose that receiver front-end consists of IQ-mixing stage at frequency ω_0 (instead of N_C mixers in standard MT CDMA receiver). IQ-mixing stage is followed by integrate-and-dump section. Continuous-time received signal is converted to discrete-time by integration over interval $T_n=T_C/Q$ where Q is oversampling factor. Equivalent complex received sequence ,$r(l)$ can be expressed as

$$r(l)=n(l)+\sum_{k=1}^{K}\sum_{i=0}^{N_c-1}\sqrt{P_k^{(i)}}\exp(j\theta_k^{(i)''})\frac{1}{T_n}\int_{(l-1)T_n}^{lT_n}s_k^{(i)}(t-\tau_k^{(i)})\exp(j(i\Delta\omega_{sc}+\Delta\omega_{Dk}^{(i)})t)dt$$

(4)

The resulting discrete-time signal is converted to a sequence of observation vectors $\boldsymbol{r}(m)\in\mathbf{C}^{QN}$ corresponding to mth bit interval which can be expressed as $\boldsymbol{r}(m)=[r(mQN+QN), ..., r(mQN+1)]^{\mathrm{T}}$. Similarly, noise vector $\boldsymbol{n}(m)\in\mathbf{C}^{QN}$ can be defined as $\boldsymbol{n}(m)=[n(mQN+QN),...., n(mQN+1)]^{\mathrm{T}}$

Bit-timing is unknown to the receiver since the system is asynchronous and an observation vector contain the end of the previous bit and the beginning of the current bit for each user. Thus, the data signals are contained in $2KN_C$-dimensional subspace of $\mathbf{C}^{QN}$.

Observation vector $\boldsymbol{r}(m)$ can be expressed in matrix form in such a way

$$\boldsymbol{r}(m)=\boldsymbol{A}\boldsymbol{c}(m)+\boldsymbol{n}(m) \tag{5}$$

Column of matrix $A\in\mathbf{C}^{QN\times2KN_C}$ correspond to signal vectors. Matrix A contain KN_C signal vectors pair of the general form: $\{u_k^{R(i)}(\tau_k^{(i)},i\Delta\omega_{sc}+\Delta\omega_{Dk}^{(i)}),u_k^{L(i)}(\tau_k^{(i)},i\Delta\omega_{sc}+\Delta\omega_{Dk}^{(i)})\}$. It is important to notice that signal vectors depend on unknown time delay, known frequency subcarrier separation and unknown frequency shift of each subcarrier which is due to subcarrier oscillator errors or Doppler shift. Problem is generally formulated so that each subcarrier has its time-delay and frequency shift. From the point of subspace based time delay and frequency shift estimation, key step is connected with modeling of signal vectors.

Vector $c(m) \in C^{2KN_C}$ depends on amplitude, phase and transmitted bits of all users and subcarriers. The estimation problem of interest can be formulated as follows: given M observation vectors $r(m)$, estimate deterministic unknown parameters $\tau_k^{(i)}$ and $\Delta\omega_{Dk}^{(i)}$, $k=1,K; i=1,N_C$ (Suppose that those parameters remain approximately constant during period of M bits).

3. SUBSPACE CONCEPT

The correlation matrix $\boldsymbol{R}$ of observation vectors can be defined and expressed in terms of its eigenvalue/eigenvector decomposition $R = E\{r(m)r(m)^*\} = VDV^*$ where * denotes conjugate transpose. The columns of matrix $V \in C^{QNxQN}$ are eigenvectrors of $\boldsymbol{R}$ and $\boldsymbol{D}$ is diagonal matrix of the corresponding eigenvalues. The $2KN_C$ largest eigenvalues correspond to the signal subspace, so the matrix V can be partitioned as $V=[E_S \; E_n]$. Columns of $E_n=[\mathbf{v}_{2KN_C+1} \dots \mathbf{v}_{QN}] \in \mathbf{C}^{QNx(QN-2KN_C)}$ form a basis for the noise subspace. Correlation matrix $\boldsymbol{R}$ is unknown and it can be estimated from M observation vectors $\boldsymbol{r(m)}$.

4. MUSIC BASED JOINT TIME-DELAY AND FREQUENCY SHIFT ESTIMATION

It is known from the theory of MUSIC algorithm [9] that signal vectors (columns of matrix A) are contained in signal subspace for the true values of $\tau_k^{(i)}$ and $\Delta\omega_{Dk}^{(i)}$. So. those signal vectors are orthogonal to the noise subspace. Form the MUSIC estimate of unknown $\tau_k^{(i)}$ and $\Delta\omega_{Dk}^{(i)}$ as

$$\{\tau_k^{(i)}, i\Delta\omega_{sc} + \Delta\omega_{Dk}^{(i)}\} = \underset{\substack{\tau\in[0,T) \\ \omega\in(-\delta,\delta+2\pi N_C/T_S)}}{\arg\max} P_{MUS}(\tau,\omega)$$

were $$P_{MUS}(\tau,\omega) = \frac{\|u_k^R(\tau,\omega)\|^2}{\|E_n^* u_k^R(\tau,\omega)\|} + \frac{\|u_k^L(\tau,\omega)\|^2}{\|E_n^* u_k^L(\tau,\omega)\|} \tag{6}$$

Unknown time delay and frequency shift parameters of the k*th* user are determined as argument of N_C maximums of $P_{MUS}(\tau,\omega)$.

5. RESULTS

In the next example we will illustrate performance of proposed algorithm for joint time-delay and frequency shift estimation applied to MT-CDMA systems. Results are given by simulation of MT-CDMA system with K=2 users and N_C=4 subcarriers per user. In example, subcarrier frequency separation is $\Delta\omega_{SC}=2\pi/T$; $\Delta\omega_{Dk}^{(i)}=0$. There is no oversampling. User's code waveforms, $b_k(t)$, and data streams , $d_k^{(i)}$, are generated using MATLAB *random* function. The length of the user's spreading sequences (which are binary random sequences) is $N=T_S/T_C=64$. The number of bits per subcarrier is M=200. Time delay for the first user is $16T_C$ and for the second $48T_C$. Time delays are equal on all subcarriers. Signal to noise ratio is 20 dB for each user. Preliminary results presented in the Fig.1 nad Fig.2. shows that maximums of $P_{MUS}(\tau,\omega)$ correspond to the subcarrier frequency separation and time delays. Performances of presented algorithm would be the subject of future study.

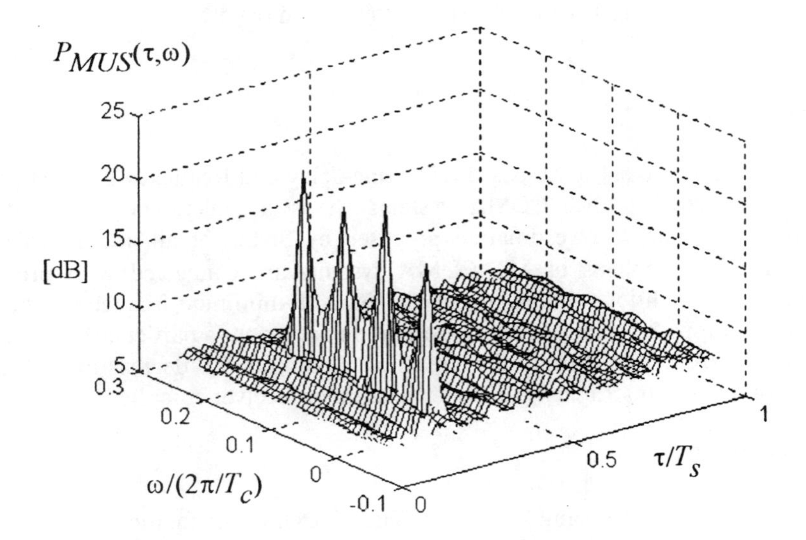

Fig 1. $P_{MUS}(\tau,\omega)$ for the first user

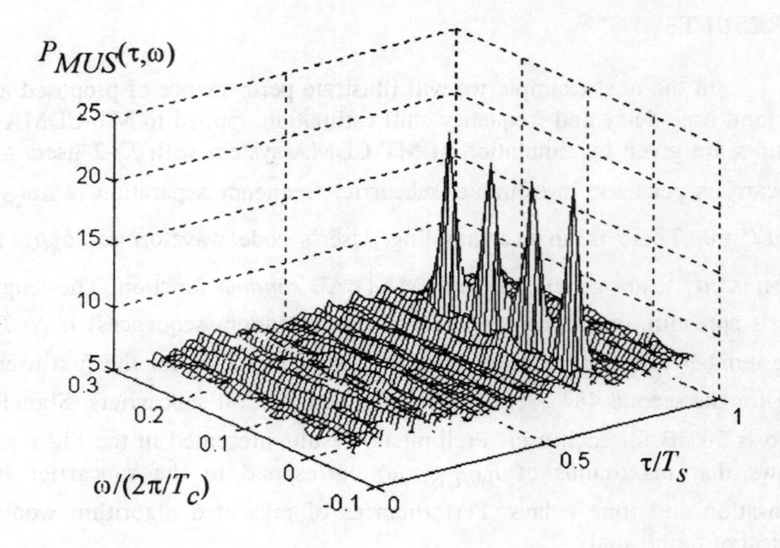

Fig 1. $P_{MUS}(\tau,\omega)$ for the second (b) user

6. CONCLUSION

We proposed subspace based joint time-delay and frequency shift estimation algorithm applied to MT-CDMA systems. It is a generalization of MUSIC algorithm for time-delay estimation proposed by Ström at all and Bensley and Aazhang. Key problems of MT CDMA systems are connected with providing orthogonality conditions signals on subcarriers with minimum frequency subcarrier separation. Proposed subspace based estimation approach is particularly interesting in MT-CDMA systems from the point of possibility of decreasing subcarrier separation, increasing capacity and optimization of receiver complexity.

7. REFERENCES

[1] S.Bensley,B.Aazhang:"Subspace-Based Delay Estimation for CDMA Communication Systems", ISIT'94, Norway.

[2] E.G. Ström :"Statistical Properties of the MUSIC Synchronization Method for DS-CDMA Systems", IEEE VTC 1996,

[3] E.G. Ström :"Subspace-Based Synchronization of DS-CDMA Systems Utilizing Training Sequences", ISSSTA'96, Mainz., Germany.

[4] E.G. Ström ,S Parkvall,S.L.Miller,B.E.Ottersten:"Propagation Delay Estimation in Asynchronous Direct- Sequence Code-Division Multiple Access Systems",IEEE Transaction on Communication, Vol.44.No1., January 1996.
[5] S.E.Bensley,B.Aazhang:"Subspace-Based Channel Estimation for Code Division Multiple Access Communication Systems",IEEE Transactions on Communications, Vol.44.No.8. August 1996.
[6] H.Liu,G.Xu:"A Subspace Method for Signature Waveform Estimation in Synchronous CDMA Systems", IEEE Transactions on Communications,Vol.44.No.10.Oct. 1996.
[7] R. Prasad,S.Hara:"An Overview of Multi-Carrier CDMA", ISSSTA'96, Mainz, Germany
[8] S.Hara,R.Prasad:"DS-CDMA,MC-CDMA and MT-CDMA for Mobile Multi-Media Communications", Proc.of IEEE VTC'96, Atlanta,USA, April 1996.
[9] R.Schmidt: "Multiple Emitter Location and Signal Parameter Estimation", IEEE Transaction on Antennas and Propagation, Vol. AP-34, No.3., March 1986.
[10] A.Polidoros, S.Glišić:"Code Synchronization:A review of Principles and Techniques",ISSSTA'94,Finland;

PILOT-SYMBOL-AIDED CHANNEL ESTIMATION IN TIME AND FREQUENCY

Peter Hoeher, Stefan Kaiser, and Patrick Robertson

Institute for Communications Technology
German Aerospace Research Establishment (DLR)
P.O. Box 1116, D-82230 Oberpfaffenhofen, Germany
E-mail: Peter.Hoeher@dlr.de

ABSTRACT

The potentials of pilot-symbol-aided channel estimation in two dimensions are explored for mobile radio and broadcasting applications. In order to procure this goal, the discrete shift-variant 2-D Wiener filter is analyzed given an arbitrary sampling grid, an arbitrary (but possibly optimized) selection of observations, and the possibility of model mismatch. Filtering in two dimensions is revealed to outperform filtering in just one dimension with respect to overhead, mean-square error performance and latency. Conceptually, the discrete shift-variant 2-D Wiener filter is the optimal linear estimator for the given problem, however, two cascaded orthogonal 1-D filters are simpler to implement and virtually as good as true 2-D filters. Analytical results are presented, verified by Monte-Carlo simulations.

1 INTRODUCTION

The basic principle of pilot-symbol-aided channel estimation is to multiplex training symbols known to the receiver into the data stream. Hence, the receiver is able to estimate the channel at any time given the observations at the pilot locations, assuming the sampling rate is sufficient with respect to the channel bandwidth[1] [1]. Pilot-symbol-aided channel estimation in the time domain (or frequency domain) is well understood [1]-[5]. Different 1-D interpolation filters have been investigated.

[1] In our definition, the channel bandwidth includes any frequency offset.

K. Fazel and G.P. Fettweis (eds.), Multi-Carrier Spread-Spectrum, 169-178.

When the channel is probed simultaneously in both time and frequency domains, the overhead of pilot symbols can be reduced significantly as proposed by the first author in [6]. Presently, state of the art is filtering with two cascaded orthogonal 1-D filters [6]-[8], referred to as $2\times$ 1-D filtering. Channel estimation in two dimensions is particularly attractive for multi-carrier modulation such as Orthogonal Frequency-Division Multiplexing (OFDM). Applications include the recent European Terrestrial Digital Video Broadcasting (DVB-T) standard [9], future mobile radio systems and multi-carrier modems. The results are directly applicable to OFDM, but with modifications also to (equalized) single-carrier systems, when training data of adjacent frequency bands is available.

Pilot-symbol-aided channel estimation provides robust estimates even for severe channel conditions and is suitable for coherent demodulation of signals with large alphabet sizes; demodulation and channel estimation (equivalent to carrier sync) are decoupled. 2-D channel estimation of an OFDM/DPSK system using training only for acquisition purposes was studied in [10].

In this paper, we investigate pilot-symbol-aided channel estimation with a single 2-D FIR filter [11], and compare the results with 1-D and $2\times$ 1-D filtering. Though this solution was briefly mentioned in [6], we could not find a detailed examination in the literature. In order to procure this goal, in Section 2 the discrete shift-variant 2-D Wiener filter is briefly derived and a mean-square error (MSE) analysis is provided. Model mismatch is included in the analysis. Realization aspects are covered in Section 3. Sections 2 and 3 are general, i.e. not restricted just to channel estimation, which is treated in Section 4 for a frequency-selective mobile radio channel. Different sampling grids (rectangular, diagonal, and random) are investigated. Given a rectangular grid, 2-D filtering is compared with 1-D filtering and $2\times$ 1-D filtering in terms of the MSE and a power-bandwidth diagram. Finally, the conclusions are drawn in Section 5.

2 SHIFT-VARIANT 2-D WIENER FILTER

Discrete (and continuous) formulations of the **shift-invariant** 2-D Wiener filter are well established in the literature [12, 14]. The **shift-variant** 2-D Wiener filter is less familiar [13]. In the context of pilot-symbol-aided channel estimation, the shift-variant 2-D Wiener filter is the optimal linear estimator in the sense of minimizing the MSE given certain constraints introduced in the following. The 2-D discrete, shift-variant minimum mean-square error (MMSE) estimation problem and its solution is as follows [13, 11]: Assume it is of interest to estimate a wide-sense stationary (WSS) 2-D stochastic process[2] $h(k,l) \in \mathbb{C}$ dis-

[2]In case of channel estimation, $h(k,l)$ is the Fourier transform of the time-variant channel impulse response $h(t,\tau)$ with respect to the delay τ ("time-frequency response") at discrete time and frequency indices.

turbed by a WSS zero-mean additive discrete noise process $n(k,l) \in \mathbb{C}$:

$$r(k,l) = h(k,l) + n(k,l) \quad \forall\, 0 \leq k \leq K-1,\ 0 \leq l \leq L-1, \tag{1}$$

where $r(k,l)$ is the received bisequence[3], $0 \leq k \leq K-1$ and $0 \leq l \leq L-1$ are the two indices of the process (e.g., time/frequency indices for the example of channel estimation), and $K \times L$ is the arbitrary (finite or infinite) block size. Assume further that $r(k,l) \in \mathbb{C}$ is observed only within a sampling grid (later: a set of pilot positions), $\mathcal{P}$. The corresponding observation indices are denoted as k' and l', respectively, where again $0 \leq k' \leq K-1$ and $0 \leq l' \leq L-1$. The number of observations (later: the number of pilot symbols) is $N_{grid} = ||\mathcal{P}|| \leq KL$. The estimator is chosen to be discrete, linear, and shift-variant:

$$\hat{h}(k,l) = \sum_{\{k',l'\} \in \mathcal{T}(k,l)} w(k',l';k,l)\, r(k',l'), \tag{2}$$

where $\hat{h}(k,l) \in \mathbb{C}$ is the estimated bisequence, $w(k',l';k,l) \in \mathbb{C}$ is the shift variant impulse response of the filter, and $\mathcal{T}(k,l) \in \mathcal{P}$ is the set of observations that are actually used. The number of filter coefficients is $N_{tap} = ||\mathcal{T}(k,l)|| \leq N_{grid}$.

The optimal linear filter in the sense of minimizing $J(\underline{w}(k,l)) = E[|h(k,l) - \hat{h}(k,l)|^2]$, i.e. the MMSE estimator or 2-D Wiener filter, (if existent) is [11]

$$\underline{w}_0^T(k,l) = \underline{\theta}^T(k,l)\, \underline{\Phi}^{-1}, \tag{3}$$

where $\underline{\Phi}$ is the N_{tap} by N_{tap} auto-covariance matrix of the channel with elements $\Phi(k'-k'', l'-l'') = E[r(k',l')\, r^*(k'',l'')]$, and $\underline{\theta}(k,l)$ is the N_{tap} by 1 cross-covariance vector with elements $\theta(k-k'', l-l'') = E[h(k,l)\, r^*(k'',l'')]$. Note that with $u = k-k'$, $v = l-l'$ and $w_o(k',l';k,l) = w_o(u,v)$ the shift-invariant solution is obtained [12, 14]. In case of $K=1$ or $L=1$, the shift-variant ("asymmetric") 1-D Wiener filter [6] is obtained. Note also that the Wiener filter is time invariant and does not depend on the actual channel state.

The MSE is [11]

$$J(\underline{w}(k,l)) = \sigma_h^2 - \underline{\theta}^T(k,l)\, \underline{w}^*(k,l) - \underline{w}^T(k,l)\, \underline{\theta}^*(k,l) + \underline{w}^T(k,l)\, \underline{\Phi}\, \underline{w}^*(k,l), \tag{4}$$

which is valid for *any* FIR filter $\underline{w}$. Model mismatch, i.e. when the covariances used in the filter design and those of the actual channel do not match, is hence included in (4). The design of the sampling grid $\mathcal{P}$, the selection of observations $\mathcal{T}(||, \updownarrow)$, and edge effects are also included in (4). The MMSE

$$J(\underline{w}_0(k,l)) = \sigma_h^2 - \underline{\theta}^T(k,l)\, \underline{\Phi}^{-1}\, \underline{\theta}^*(k,l) \leq J(\underline{w}(k,l)) \tag{5}$$

is approached for known covariances (no mismatch).

In [11] we have calculated the covariances for Bello's WSSUS channel model [16], which fits our assumptions well.

[3] In the situation of pilot-symbol-aided channel estimation, $r(k,l)$ is obtained by dividing the matched filter outputs through the training symbols, i.e., by re-modulation.

3 REALIZATION ASPECTS

Realization aspects (e.g., existence, stability, pre-computation of filter sets, etc.) are similar to the 1-D [1]-[5] or 2× 1-D [6]-[8] cases, see [11]. Of particular interest here, however, is a) the grid design (we investigated different sampling grids $\mathcal{P}$), and b) the selection of $\mathcal{T}(k,l) \in \mathcal{P}$. Given $N_{tap} < N_{grid}$, the optimal selection criteria is to check for each location $\{k,l\}$ all possible constellations $\mathcal{T}(k,l)$ and to select that constellation which minimizes $J(\underline{w}(k,l))$. Since this optimum rule is rather complex for practical block sizes, the sampling positions "nearest" to the actual location $\{k,l\}$ might be selected instead.

The latency generally depends on the number of filter coefficients, N_{tap}, the design of the sampling grid, and the selection criteria. Exploiting the correlation in the frequency domain should be used to reduce the latency.

4 PILOT-SYMBOL CHANNEL ESTIMATION

4.1 GRID DESIGN

For mobile radio and broadcasting applications, let us consider a time invariant grid design, i.e., which is independent of the actual channel state. Let us denote the spacing between pilot symbols in the time domain as N_K and in the frequency domain as N_L, the one-sided maximum Doppler frequency as $f_{D_{max}}$, the one-sided maximum echo delay as τ_{max}, the symbol duration as T_s and the carrier spacing as ΔF.

Given the normalized channel bandwidths $f_{D_{max}} T_s$ and $\tau_{max}\Delta F$, the sampling theorem [15] requires that

$$f_{D_{max}} T_s \cdot N_K \leq 1/2 \quad \text{and} \quad \tau_{max}\Delta F \cdot N_L \leq 1/2. \tag{6}$$

We define a "balanced design" to be

$$f_{D_{max}} T_s \cdot N_K \approx \tau_{max}\Delta F \cdot N_L. \tag{7}$$

The balanced design is optimal if the Doppler power spectrum and the delay power spectrum would have the same shape (for example, if they would be rectangular); the proof is clear by symmetry arguments.

A "rule of thumb" is 2× oversampling to achieve reasonable complexity (filter length) and performance. In this case

$$f_{D_{max}} T_s \cdot N_K \approx 1/4 \quad \text{and} \quad \tau_{max}\Delta F \cdot N_L \approx 1/4. \tag{8}$$

Other sampling rates are studied in Section 4.4.

Example: Consider $f_{D_{max}} = 320$ Hz (e.g., $v = 192$ km/h @ $f_0 = 1.8$ GHz) and $\tau_{max} = 10\mu$s (i.e., 20μs maximum delay) as worst-case parameters of a typical land mobile radio system.

Given these constraints, a balanced design with 2x oversampling applied to an OFDM system is as follows:

$$\begin{aligned} f_{D_{max}} T_s N_K &= 320 \text{ Hz} \cdot 260\mu\text{s} \cdot 3 = 1/4 \\ \tau_{max} \Delta F N_L &= 10\mu\text{s} \cdot 4.1667 \text{ kHz} \cdot 6 = 1/4, \end{aligned} \tag{9}$$

where $N_K = 3$, $N_L = 6$. The useful symbol duration is $T = T_s - \Delta = 260\ \mu\text{s} - 20\ \mu\text{s} = 240\ \mu\text{s}$, the guard interval is $\Delta = 2\tau_{max} = 20\ \mu$s, and the carrier spacing is $1/T = \Delta F = 4.1667$ kHz.

This scheme was investigated for short blocks with $K = 58$ and $L = 115$. A rectangular sampling grid with $N_K = 3$ and $N_L = 6$ consists of $N_{grid} = 20 \times 20 = 400$ pilot symbols; the overhead is $1/(N_K N_L)$ excluding edge effects. Other grids (diagonal and random) were examined as well.

4.2 MMSE PERFORMANCE

Performance criterion is the MSE averaged over all symbol positions (edge effects are hence included) versus E_s/N_0, where E_s/N_0 is the SNR of the training symbols. In Section 4.2.1-4.2.3, perfect knowledge of the covariances is assumed to avoid any spurious effects. Model mismatch is then considered in Section 4.3. Analytical results (represented by lines) were verified by simulations (represented by symbols). Note again that the channel conditions are severe.

4.2.1 1-D CHANNEL ESTIMATION

1-D channel estimation serves as a benchmark. Since $N_L > N_K$ we looked at filtering in the frequency domain; the overhead is $1/N_L$ excluding edge effects. The results are reported in Fig. 1 given various numbers of coefficients, N_{tap}. About five filter coefficients are sufficient. Then, for example, a MSE of 10^{-2} is obtained at about 18.8 dB. For comparison, an ideal low-pass interpolator is considered in Fig. 1 as well. By construction, 1-D filters operating in the frequency domain are unable to make use of temporal correlations of the channel.

4.2.2 2× 1-D CHANNEL ESTIMATION

The second strategy under investigation is 2× 1-D channel estimation [6]: filtering in the frequency domain is followed by filtering in the time domain, both using 1-D Wiener filters. The actual order of filtering (time or frequency first)

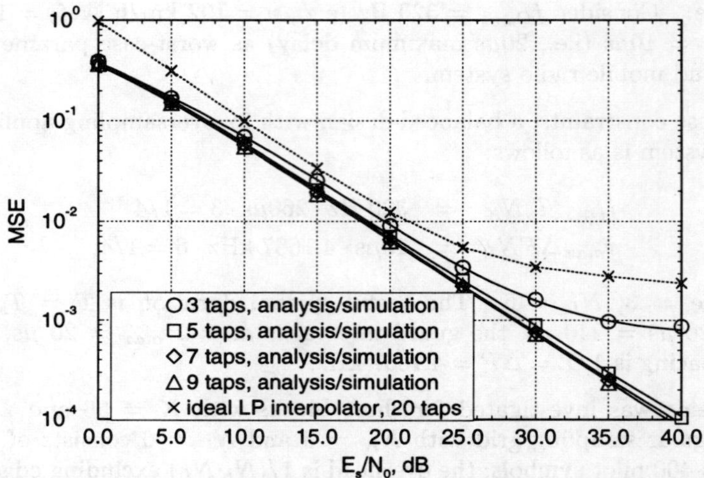

Figure 1: Average MSE versus SNR applying 1-D Wiener filtering. (No mismatch. $N_L = 6$, $\tau_{max}\Delta F N_L = 0.25$, $L = 115$. For comparison, the MSE for an ideal low-pass interpolator is also included.)

is arbitrary due to linearity. Note that the noise process seen by the second filter remains white, but its SNR is improved due to the first filter. This SNR improvement has not been considered in the design of the second filter in earlier publications [6, 7, 8].

Given the same overhead as above, the improvement compared to 1-D filtering for practical SNR is pronounced: with 2×5 taps a MSE of 10^{-2} is obtained at about 15.0 dB. For high SNR, however, 1-D filtering approaches the performance of $2\times$ 1-D filtering. About 2×5 taps are sufficient.

From now on we set $N_K = 3$ and $N_L = 6$, i.e. we reduce the overhead by a factor of three. A MSE of 10^{-2} is obtained at about 17.3 dB. The MSE is still lower than for 1-D filtering, although the total pilot overhead is reduced.

4.2.3 2-D CHANNEL ESTIMATION

Finally, channel estimation with a single 2-D Wiener filter is investigated. 10, 25 and 100 taps were chosen: Comparing the complexity, 10 taps (2-D) corresponds to 2×5 taps ($2\times$ 1-D), 25 taps corresponds to 5×5 taps, and 100 taps serves a benchmark.

Let us start with the rectangular grid. The results are reported in Fig. 2. If $N_{tap} = 10 = 2 \times 5$, the MSE is similar to the MSE of $2\times$ 1-D filtering for practical SNR: a MSE of 10^{-2} is obtained at about 17.8 dB. If $N_{tap} = 25 = 5\times5$, the MSE of 2-D filtering is always better: 10^{-2} is obtained at about 16.7 dB.

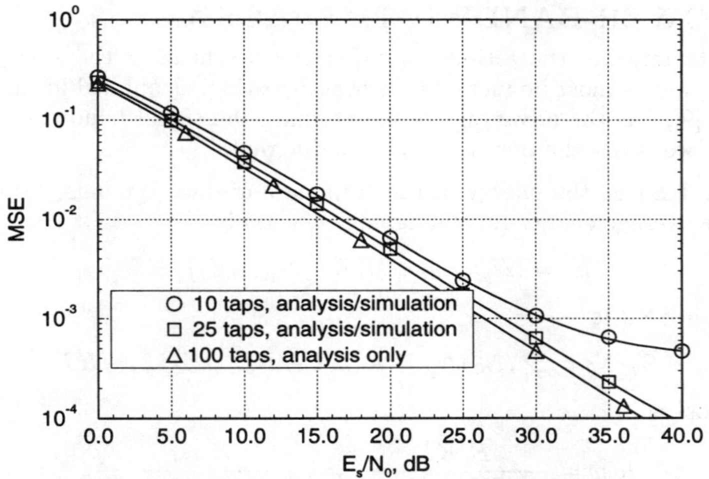

Figure 2: Average MSE versus SNR applying 2-D Wiener filtering given a rectangular grid. (No mismatch. $N_K = 3$, $N_L = 6$, $f_{D_{max}} T_s N_K = 0.25$, $\tau_{max} \Delta F N_L = 0.25$, $K = 58$, $L = 115$.)

The MSE results for a diagonal grid are comparable when averaged over all symbol positions. The edges are less well probed compared to a rectangular grid. The diagonal grid can also been viewed as a (stretched) hexagonal grid.

Finally, random grids were studied. The MSE was averaged over 20 different random grids with $N_{grid} = 400$. Because the sampling theorem is violated for some positions $\{k, l\}$, the results are much worse.

4.3 PERFORMANCE WITH MODEL MISMATCH

Wiener interpolation offers optimal (MMSE) performance by using knowledge about the channel statistics. This generally requires an adaptation. A certain degradation occurs if the covariances of the channel and of the filter differ, c.f. (4) and (5).

Another philosophy of robust, low effort interpolation is the use of a fixed FIR filter matched to the worst case channel conditions. In order to investigate the influence of model mismatch in this case, we used a filter designed for a rectangular Doppler/delay spectrum (- with disclaimers about optimality of this filter -) and tested different channel models defined by COST 207 (typical urban, bad urban, and hilly terrain). In accordance to the 1-D [3, 4] and 2× 1-D cases [7], model mismatch does not seriously degrade the estimation performance as long as the filter bandwidths exceed the channel bandwidths.

4.4 POWER-BANDWIDTH DIAGRAM

In order to improve the estimates, either the overhead or the energy of the training symbols must be increased. Power boosting, as applied in the DVB-T standard [9], has the advantage of maintaining the signal bandwidth. In the following, we study the power-bandwidth diagram.

Let E_s and E_d be the energy of the training and data symbols, respectively. Then, the average energy (including edge effects) is

$$\overline{E}_s = (N_{grid}E_s + (KL - N_{grid})E_d)/KL \tag{10}$$

the average SNR is

$$\overline{E}_s/N_0 = E_s/N_0 \left(N_{grid} + (KL - N_{grid})E_d/E_s\right)/KL \tag{11}$$

and the loss is

$$loss = \frac{\overline{E}_s KL}{E_d(KL - N_{grid})} = 1 + \frac{N_{grid}}{KL - N_{grid}}\frac{E_s}{E_d}. \tag{12}$$

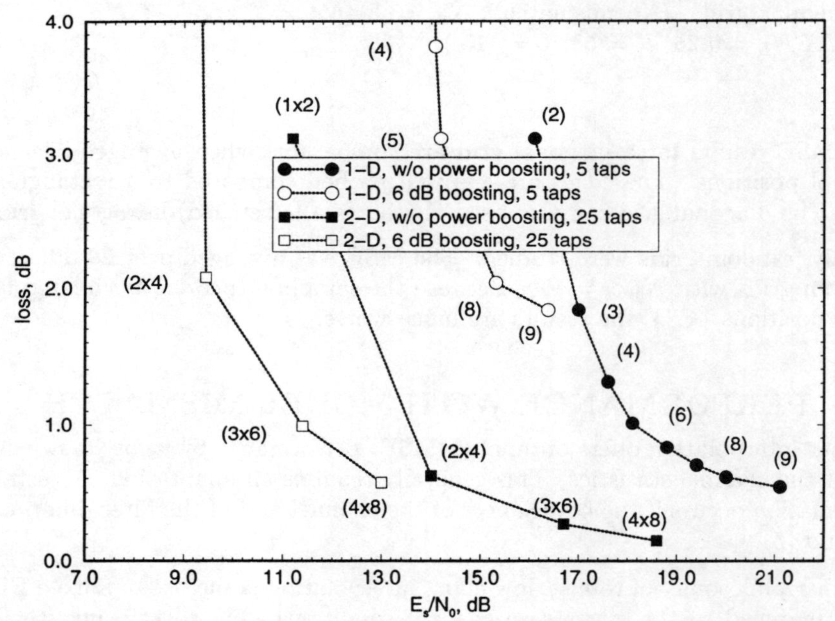

Figure 3: Loss vs. SNR @ MSE=10^{-2} applying 1-D and 2-D Wiener filtering. (No mismatch. Rectangular grid. N_K and N_L are denoted in brackets. $f_{D_{max}} T_s N_K = 0.25$ for $N_K = 3$, $\tau_{max}\Delta F N_L = 0.25$ for $N_L = 6$.)

The goal is now to acquire a certain MSE. In Fig. 3 we have plotted the loss versus the necessary $\overline{E}_s/N_0$ given an average MSE of 10^{-2} for a power boosting, E_s/E_d, of 0 dB and 6 dB, respectively. Both the 1-D and 2-D cases are considered. In the latter situation a balanced design was applied. A rectangular sampling grid was chosen. The numbers in brackets indicate the effective spacing, $N_K \times N_L$ (1-D: N_L). The block size was adjusted so that the grid consists of $N_{grid} = 20 \times 20 = 400$ pilot symbols (1-D: 20 pilot symbols). Edge effects were taken into account by averaging the MSE over all symbol positions. Fig. 3 illustrates that 2-D channel estimation significantly outperforms 1-D channel estimation. In particular, power boosting is very efficient in the 2-D case due to the lower redundancy.

When plotting the overhead (instead of the loss) versus $\overline{E}_s/N_0$, we would obtain the corresponding power-bandwidth diagram.

5 CONCLUSIONS

The potentials of pilot-symbol-aided channel estimation in one and two dimensions were explored. The discrete shift-variant 2-D Wiener filter was briefly derived and analysed. The MSE analysis is valid for an arbitrary sampling grid, an arbitrary selection of observation points and filter order, the possibility of model mismatch and holds for any FIR filter.

Our main results are as follows:

- Filtering in two dimensions (2-D or 2× 1-D) outperforms filtering in just one dimension with respect to overhead, MSE performance, and latency. Channel estimation is possible even for moderately fast fading and long echo delays. The overhead, a principle drawback of pilot-aided schemes, can be reduced significantly. Hence, the power *and* the bandwidth efficiencies are improved.
- The performance of 2× 1-D filtering is similar to 2-D filtering, but 1-D filters are possibly simpler to implement.
- Model mismatch (i.e., $\underline{w}(k,l) \neq \underline{w}_0(k,l)$) is of minor influence, as long as the filter bandwidths exceed the channel bandwidths in both dimensions. In practice, it might be sufficient to design the (fixed) filter coefficients for the worst-case channel conditions.
- Given (4), a grid design is possible, i.e., communication systems can be optimized with respect to channel estimation.
- Best performance is achieved by a "balanced design" (same effective sampling ratio in time and frequency domains).
- A rectangular and a diagonal grid lead to a similar performance, both for the balanced design under examination within this paper and further unbalanced scenarios where the Doppler spread (respectively the delay spread) was assumed to be small. A random grid is much worse.
- Power boosting is efficient when the overhead is small (i.e., in the 2-D case). It maintains the bandwidth efficiency. Especially edge effects could reduced by power boosting.

References

[1] M.L. Moher and J.H. Lodge, "A time diversity modulation strategy for the satellite-mobile channel," in *Proc.* 13th *Biennial Symp. Commun.*, Queen's Univ. Kingston, Canada, June 1986.

[2] A. Aghamohammadi, H. Meyr, and G. Ascheid, "A new method for phase synchronization and automatic gain control of linearly modulated signals on frequency-flat fading channels," *IEEE Trans. Commun.*, vol. COM-39, pp. 25-29, Jan. 1991.

[3] J.K. Cavers, "An analysis of pilot symbol assisted modulation for Rayleigh fading channels," *IEEE Trans. Vehicular Techn.*, vol. VT-40, pp. 686-693, Nov. 1991.

[4] S. Fechtel, "Verfahren und Algorithmen der robusten Synchronisation für die Datenübertragung über dispersive Schwundkanäle," Ph.D. thesis, RWTH Aachen, Germany, 1993.

[5] P. Schramm, "Modulationsverfahren für CDMA-Mobilkommunikationssysteme unter Berücksichtigung von Kanalcodierung und Kanalschätzung," Ph.D. thesis, University of Erlangen-Nuremberg, Germany, 1996.

[6] P. Hoeher, "TCM on frequency-selective land-mobile fading channels," in *Coded Modulation and Bandwidth-Efficient Transmission.* Elsevier Science Publishers B.V., E. Biglieri and M. Luise (eds.), pp. 317-328, 1992.

[7] F. Classen, M. Speth, and H. Meyr, "Channel estimation units for an OFDM system suitable for mobile communication," in *Proc. ITG Fachtagung Mobile Kommunikation*, Neu-Ulm, Germany, Sept. 1995.

[8] T. Mueller, K. Brueninghaus, and H. Rohling, "Performance of coherent OFDM-CDMA for broadband mobile communications," *Wireless Personal Communications*, Kluver Academic Publishers, vol. 2, pp. 295-305, 1996.

[9] ETS 300 744 Standard, "Digital broadcasting systems for television, sound, and data services; Frame structure, channel coding and modulation for digital terrestrial broadcasting," *ETSI*, 1996.

[10] P. Frenger and A. Svensson, "A decision directed coherent detector for OFDM," in *Proc. Vehicular Techn. Conf.* '96, Atlanta, Georgia, pp. 1584-1588, Apr.-May 1996.

[11] P. Hoeher, S. Kaiser, and P. Robertson, "Two-dimensional pilot-symbol-aided channel estimation by Wiener filtering," in *Proc. IEEE ICASSP* '97, Munich, Germany, pp. 1845-1848, Apr. 1997.

[12] C.W. Helstrom, "Image restoration by the method of least squares," *J. Opt. Soc. Amer.*, vol. 57, pp. 297-303, Mar. 1967.

[13] W.K. Pratt, "Generalized Wiener filtering computation techniques," *IEEE Trans. Computers*, vol. C-21, no. 7, pp. 636-641, July 1972.

[14] M.P. Ekstrom, "Realizable Wiener filtering in two dimensions," *IEEE Trans. Acoust., Speech, Signal Proc.*, vol. ASSP-30, no. 2, pp. 31-40, Feb. 1982.

[15] R.M. Mersereau and T.C. Speake, "The processing of periodically sampled multidimensional signals," *IEEE Trans. Acoust., Speech, Signal Proc.*, vol. ASSP-31, no. 2, pp. 188-194, Feb. 1983.

[16] P.A. Bello, "Characterization of randomly time variant linear channels," *IEEE Trans. Commun. Syst.*, vol. CS-11, pp. 360-393, Dec. 1963.

A Family of Extended Gaussian Functions with a Nearly Optimal Localization Property

Christian Roche and Pierre Siohan

DSM/RSA, Laboratory of CCETT - France Telecom Group
Rue du Clos Courtel, 35512 Cesson Sévigné Cedex, FRANCE,
e-mail : siohan@ccett.fr

Abstract

Localization in time and frequency is a key issue in the design of multicarrier modulation systems. A new prototype function has been recently proposed to build an OFDM/OQAM modulation scheme which nearly succeeds to reach the optimal localization value. This paper gives a closed-form expression for this prototype function and studies its localization property with respect to its shape parameter.

1 Introduction

With the European Digital Audio Broadcasting (DAB) project, Orthogonal Frequency Division Multiplexing (OFDM) has proven to be a key technique for digital terrestrial broadcasting. For this multicarrier system, as well as the one proposed later on for digital television broadcasting, the usual option is a rectangular time-window shaping, including a guard interval in order to protect the transmitted signal against echoes. Another approach has been proposed

K. Fazel and G.P. Fettweis (eds.), Multi-Carrier Spread-Spectrum, 179-186.

more recently [1], [2], where it is suggested that, in order to face the time and frequency distortions due to a dispersive transmission channel, as well as other impairments due to the receiver implementation (e.g. local oscillator offset and phase noise), the basic signals must be optimally localized with regards time and frequency. As a result, a new transform, named IOTA (Isotropic Orthogonal Transform Algorithm), was derived by M. Alard [1]. This transform which is orthogonal and nearly isotropic almost manages to fulfil the two incompatible requirements of optimal localization and orthogonality. Unfortunately, in these latter references, the so-called IOTA prototype function, denoted z_α in [2], is obtained by an algorithmic construction and, furthermore its analysis is limited to a value of α equal to 1. Our aim in this paper is to give an analytical expression for z_α, thus making this transform more tractable and simplifying its analysis and use in multicarrier systems. Section II of our paper gives an overview of the principles and properties behind the new time-frequency shaping concept named IOTA. Section III presents the mathematical derivation leading to a closed-form expression of IOTA. This analytical expression is used in Section IV to analyze the localization properties of the IOTA modulation scheme with respect to the parameter α.

2 OFDM/QAM and IOTA modulation schemes

In this section we, firstly, review some basic principles of multicarrier systems and, secondly, we summarize some of the key elements which are proposed in [2] to build a new modulation scheme. As it is explained in a large number of publications, and one can refer for instance to [3], in standard OFDM multicarrier systems each carrier may be modulated using for instance Quadrature Amplitude Modulation (QAM). Such an ODFM/QAM signal, $s(t)$, can be written as

$$s(t) = \sum_{m,n} a_{m,n} x_{m,n}(t), \tag{1}$$

where the coefficients $a_{m,n}$ are complex numbers representing the transmitted encoded data and $x_{m,n}(t)$ represents the generic function of the signal basis. Each $x_{m,n}(t)$ function is a translated version, in time and frequency, of a prototype function $x(t)$

$$x_{m,n}(t) = e^{2j\pi m\nu_0 t} x(t - n\tau_0), \text{ with } \nu_0\tau_0 = 1 \tag{2}$$

and

$$x(t) = \begin{cases} \frac{1}{\sqrt{\tau_0}} & \text{if } |t| \leq \frac{\tau_0}{2} \\ 0 & \text{elsewhere} \end{cases}. \tag{3}$$

τ_0 is the symbol duration and ν_0 denotes the spacing between two successive carriers. In the time-frequency plane this modulation scheme leads to a lattice with density equal to 1, i.e. $\tau_0\nu_0 = 1$.

Based on the analysis provided in [1] we consider that the robustness of an OFDM modulation scheme, with respect to multipath propagation, can be estimated through the value of the parameter $\xi = \frac{1}{4\pi\Delta t\Delta\nu}$, where Δt^2 and $\Delta\nu^2$ are the second order moments of a function $x(t)$, with Fourier transform $X(\nu)$, i.e.

$$\left\{ \begin{array}{l} \Delta t^2 = \frac{1}{E}\int t^2\|x(t)\|^2 dt \\ \Delta\nu^2 = \frac{1}{E}\int \nu^2\|X(\nu)\|^2 d\nu \\ E = \int \|X(\nu)\|^2 d\nu \end{array} \right. . \tag{4}$$

Using the Heisenberg inequality it can be proved that ξ is always less than unity. In the case of an OFDM/QAM modulation scheme $\xi = 0$. This is, naturally, a major drawback making OFDM/QAM very sensitive to multipath propagation. Nevertheless if, as in the DAB system, we spread the rectangular window by means of the so-called guard interval the problem is partly solved, at a price depending on the spectral efficiency required. If for low spectral efficiency systems the corresponding loss of power may be acceptable, as for instance the DAB with only 1 dB of loss, it may attain 4 dB if a 4 bits/s/Hz spectral efficiency is required, while keeping a similar guard interval as for the DAB, i.e. one quarter of the useful symbol.

To get rid of the guard interval a possible choice lies in an Offset Quadrature Amplitude Modulation of each carrier. OFDM/OQAM has also been studied by many researchers and [4] is one of the corresponding references. The OFDM/OQAM signal, $s(t)$, is also derived from relation (1), the $a_{m,n}$ coefficients now being real numbers and the basis functions being generated by

$$x_{m,n}(t) = j^{m+n} e^{2j\pi m\nu_0 t} x(t - n\tau_0), \tag{5}$$

with $\nu_0\tau_0 = 1/2$. The $x_{m,n}(t)$ functions are orthogonal and the corresponding time-frequency lattice of this modulation scheme is of density 2. A usual choice for the prototype function $x(t)$ is to derive it as the impulse response of a half-Nyquist filter. At a price of a higher implementation cost, a prefiltering being required, such an OFDM/OQAM technique may lead to an acceptable value of the time-frequency localization parameter ($\xi = 0.865$) but, unfortunately, with a poor localization in time [2].

More recently another modulation scheme has been proposed which, naturally, leads to high values of the parameter ξ. The mathematical tools used by M. Alard [1], [2] to build an orthogonal system and check its orthogonality are

- An orthogonalization operator, denoted $\mathbf{O}_a$, which transforms a function x ($\mathbb{R} \rightarrow \mathbb{R}$) into a function y by

$$y(u) = \frac{x(u)}{\sqrt{a\sum_k \|x(u - ka)\|^2}}, \tag{6}$$

- The Fourier transform operator $\mathbf{F}$,

- The Woodward ambiguity function which for a function y is given by

$$A_y(\tau,\nu) = \int y(t+\tau/2)y^*(t-\tau/2)e^{-2j\pi\nu t}dt. \tag{7}$$

As shown in [2], [1], $\mathbf{O}_{\nu_0}$ allows an orthogonalization in time while $\mathbf{F}^{-1}\mathbf{O}_{\tau_0}\mathbf{F}$ is used for the orthogonalization along the frequency axis.

Furthermore, when applied to a gaussian function, the operators $\mathbf{O}_{\nu_0}$ and $\mathbf{F}^{-1}\mathbf{O}_{\tau_0}\mathbf{F}$ commute. This leads to the definition of a new prototype function $z_\alpha = \mathbf{O}_{\nu_0}\mathbf{F}^{-1}\mathbf{O}_{\tau_0}\mathbf{F}g_\alpha$ with $g_\alpha(u) = (2\alpha)^{1/4}e^{-\pi\alpha u^2}$. It can be then proved that

$$A_{z_\alpha}(2n\tau_0, 2m\nu_0) = \begin{cases} 0 & \text{if} \quad (m,n) \neq (0,0) \\ 1 & \text{if} \quad (m,n) = (0,0) \end{cases} . \tag{8}$$

Consequently the orthonormality property is perfectly satisfied on a regular lattice. Furthermore $\mathbf{F}z_\alpha = z_{1/\alpha}$. Thus if $\alpha = 1$ we get a function identical to its Fourier transform, this function is denoted by the symbol $\mathcal{J}$. This latter function, which is at the origin of the IOTA acronym, is conjectured in [2] to be the optimum as regards to the localization property. Indeed in this case $\xi = 0.977$, and the theoretical limit of 1 corresponds to a gaussian prototype and consequently does not satisfy the orthogonality constraint.

It is probably why the analysis reported in [1], [2] is uniquely devoted to the function $\mathcal{J}$. Our own conviction is that different values of α may be suitable for different transmission channels and that some of the multiple solutions allowed by the construction method described in the patent [1] deserve specific studies. Let us now examine how a closed-form expression can be derived for z_α and which range of α values are of interest with regard to the localization property.

3 z_α : Closed-form expression

If an exhaustive analysis with respect to α is required, the original definition of the function z_α may involve long computation time. In this section we derive an analytical expression which makes the function z_α more tractable. Using the normalization proposed in [1] we set $\tau_0 = \nu_0 = \frac{1}{\sqrt{2}}$ and consequently $\mathbf{O}_{\tau_0} = \mathbf{O}_{\nu_0} = \mathbf{O}$. Therefore the prototype function $x(t)$ of equation (5) is given by $x(t) = z_\alpha(t)$. For different values of τ_0, $x(t)$ is defined as [1]

$$x(t) = \frac{1}{2^{1/4}\sqrt{\tau_0}} z_\alpha\left(\frac{t}{\tau_0\sqrt{2}}\right). \tag{9}$$

Theorem – *Let* $z_\alpha = \boldsymbol{OF}^{-1}\boldsymbol{OF}g_\alpha$ *with* $\boldsymbol{F}$ *the Fourier transform operator and* $\boldsymbol{O}$ *an orthogonalization operator (6), then*

$$z_\alpha(t) = \frac{1}{2}\sum_{k=0}^{\infty} d_{k,\alpha}[g_\alpha(t+k\sqrt{2}) + g_\alpha(t-k\sqrt{2})]\sum_{l=0}^{\infty} d_{l,1/\alpha} cos(2\pi l t\sqrt{2}), \tag{10}$$

with $\{d_{k,\alpha}\}$ a set of real coefficients and $\alpha_M \leq \alpha \leq 1/\alpha_M$ with $\alpha_M \cong 0.264$.

Proof . The proof directly uses the definition of z_α and, as a matter of consequence, can be decomposed in two main steps. Let $\mathbf{F}g_{\alpha,i} = G_{\alpha,i}$, with $0 \leq i \leq 1$ and $g_{\alpha,0} = g_\alpha$, $G_{\alpha,0} = G_\alpha$.

- **Step 1-** Computation of $\mathbf{F}^{-1}\mathbf{O}\mathbf{F}g_\alpha = \mathbf{F}^{-1}\mathbf{O}G_\alpha = \mathbf{F}^{-1}G_{\alpha,1} = g_{\alpha,1}$.

  $G_{\alpha,1}(\nu)$ is obtained by relation (6), its denominator $\sum_k G_\alpha^2(\nu - k/\sqrt{2})$ being a periodic function, with period $1/\sqrt{2}$, it can be expanded in a Fourier series

$$\sum_k G_\alpha^2(\nu - \frac{k}{\sqrt{2}}) = \sqrt{2}(1 + 2\sum_{p=1}^{\infty} e^{-\pi\alpha p^2} \cos(2\pi p\nu\sqrt{2})). \tag{11}$$

  Let $X = 2\sum_{p=1}^{\infty} e^{-\pi\alpha p^2} \cos(2\pi p\nu\sqrt{2})$, $G_{\alpha,1}(\nu)$ is then given by

$$G_{\alpha,1}(\nu) = \frac{G_\alpha(\nu)}{\sqrt{1+X}}. \tag{12}$$

  It can be checked that, as shown in [5], $|X| < 1$ if $\alpha > \alpha_M$, with $\alpha_M \cong 0.264$. Assuming this latter condition is satisfied, $G_{\alpha,1}(\nu)$ can then be expanded as

$$G_{\alpha,1}(\nu) = G_\alpha(\nu)\left[1 + \sum_{k=1}^{\infty}(-1)^k \frac{(2k-1)!}{2^{k-1}k!(k-1)!}(\sum_{p=1}^{\infty} e^{-\pi\alpha p^2} \cos(2\pi p\nu\sqrt{2}))^k\right]. \tag{13}$$

  Let

$$G_{\alpha,1}(\nu) \equiv G_\alpha(\nu)\sum_{k=0}^{\infty} d_{k,\alpha} \cos(2\pi k\nu\sqrt{2}). \tag{14}$$

  By identification, using (13), (14), we can derive, the coefficients $d_{k,\alpha}$ given below (cf. (21)), while an inverse Fourier transform of expression (14) gives

$$g_{\alpha,1}(t) = \frac{1}{2}\sum_{k=0}^{\infty} d_{k,\alpha}\left[g_\alpha(t + k\sqrt{2}) + g_\alpha(t - k\sqrt{2})\right]. \tag{15}$$

- **Step 2-** Orthogonalization of $g_{\alpha,1}(t)$.

  $\mathbf{O}g_{\alpha,1}$ is also defined by relation (6). Let $S_\alpha(t) = \sum_m g_{\alpha,1}^2(t - m/\sqrt{2})$. $S_\alpha(t)$ is a periodic function which can be written as

$$S_\alpha(t) = e_0 + 2\sum_{p=1}^{\infty} e_p \cos(2\pi pt\sqrt{2}). \tag{16}$$

After some algebraic manipulations reported in [5] we find that

$$S_\alpha(t) = \sqrt{2}D(\alpha)(1+Y) \text{ with } Y = 2\sum_{p=1}^{\infty} e^{-\frac{\pi}{\alpha}p^2} \cos(2\pi pt\sqrt{2}) \tag{17}$$

and

$$D(\alpha) = \frac{1}{2}\sum_{i=0}^{\infty}\sum_{k=0}^{\infty} d_{i,\alpha}d_{k,\alpha}\left[e^{-\pi\alpha(i+k)^2} + e^{-\pi\alpha(i-k)^2}\right]. \tag{18}$$

From what we indicated above concerning X it is clear that, similarly, $|Y| < 1$ if $\alpha < 1/\alpha_M$, i.e. if $\alpha < 3.784$. Therefore, assuming now that $\alpha_M < \alpha < 1/\alpha_M$ we can write

$$z_\alpha(t) = \frac{g_{\alpha,1}(t)}{\sqrt{D(\alpha)}}\left[1 + \sum_{k=1}^{\infty}(-1)^k \frac{(2k-1)!}{2^{k-1}k!(k-1)!}\left(\sum_{p=1}^{\infty} e^{-\frac{\pi}{\alpha}p^2}\cos(2\pi pt\sqrt{2})\right)^k\right]. \tag{19}$$

The second factor in (19) can be identified with $\sum_{l=0}^{\infty} c_{l,\alpha}\cos(2\pi lt\sqrt{2})$. By analogy with relations (13), (14) we easily deduced that $c_{l,\alpha} = d_{l,1/\alpha}$. Furthermore the expression of $z_\alpha(t)$ given by (19) may be simplified. Indeed using the fact, proved in [1], that $\mathbf{F}z_\alpha = z_{1/\alpha}$ we get [5]

$$D(\alpha) = D(\frac{1}{\alpha}) = 1, \text{ for } 0.264 < \alpha < 3.784. \tag{20}$$

That achieves the proof of relation (10).

The practical use of the expression z_α (10), naturally, requires the knowledge of the infinite set of coefficients $\{d_{k,\alpha}\}$. In fact, if no explicit expression has been found for these coefficients, we are nevertheless able to show they can be written as an infinite series such that

$$d_{k,\alpha} = \sum_{l=0}^{\infty} a_{k,l}e^{-l\pi\alpha}. \tag{21}$$

It is worthwhile noting that, owing to the fast decreasing property of z_α, a reduced number of coefficients $a_{k,l}$ are sufficient to gain a very accurate approximation. Let $\{\tilde{d}_{k,\alpha}\}$ be a finite set of coefficients corresponding to a truncated version of the set $\{d_{k,\alpha}\}$. As a matter of example, in Table 1 a set of $\{a_{k,l}\}$ coefficients is provided which can be used to compute the seven first coefficients $\tilde{d}_{k,\alpha}$ with an accuracy which is around $e^{-6\alpha\pi}$.

4 z_α : Time-frequency localization

The fast decreasing property of the function z_α, at least for $\alpha \geq 1$, is clearly demonstrated in Fig. 1 where typical curves of the functions z_α and $\mathbf{F}z_\alpha$ are plotted for $\alpha = 1/2, 1$ and 2. For this family of orthogonal Extended Gaussian Functions (EGF) the localization parameter ξ clearly depends upon the parameter α. Using the expressions of the second order moments given in equation (4) we obtain the result plotted in Fig. 2. As a matter of comparison we took as a reference the measure $\xi = 0.865$, which is obtained with the OFDM/OQAM modulation and also with the OFDM/MSK modulation (cf. [2]). It can be clearly seen in Fig. 2 that all the values of α in the range $[1/2, 2]$ allow different trade-offs in the time-frequency domain, which are all, from the localization point of view, better than other already known multicarrier schemes.

5 Conclusion

After reviewing the basic construction principles of a new multicarrier modulation technique, IOTA, a mathematical derivation has been carried out leading to a family of extended gaussian functions named EGF. This new mathematical description has been used afterwards to analyze the EGF localization property with respect to the shape parameter (α). Further work will, experimentally, investigate the role of this parameter in a complete transmission chain.

Acknowledgements : The authors would like to thank their colleagues D. Castelain, P. Combelles and D. Lacroix from RSA/DHN department for fruitful discussions concerning the OFDM/IOTA modulation system.

References

[1] M. Alard. Construction d'un signal multiporteuse. Patent WO 96/35278, 1996.

[2] B. Le Floch, M. Alard, and C. Berrou. Coded orthogonal frequency division multiplex. *Proceedings of the IEEE*, 83:982–996, June 1995.

[3] L. J. Cimini. Analysis and simulation of a digital mobile channel using orthogonal frequency division multiplexing. *IEEE Trans. on Communications*, 33:665–675, July 1985.

[4] B. Hirosaki. An orthogonally multiplexed QAM system using the discrete Fourier transform. *IEEE Trans. on Communications*, 29:982–989, July 1981.

[5] C. Roche and P. Siohan. Étude d'une famille de fonctions gaussiennes étendues et analyse de leurs propriétés de localisation temps-fréquence. Technical report in preparation.

	l (0 to 6)						
	1	0	3/4	0	105/64	0	675/256
k	0	-1	0	-15/8	0	-219/64	0
	0	0	3/4	0	19/16	0	1545/512
0	0	0	0	-5/8	0	-123/128	0
to	0	0	0	0	35/64	0	213/256
6	0	0	0	0	0	-63/128	0
	0	0	0	0	0	0	231/512

Table 1: Set of $a_{k,l}$ coefficients for a computation of $\tilde{d}_{k,\alpha}$ with seven terms.

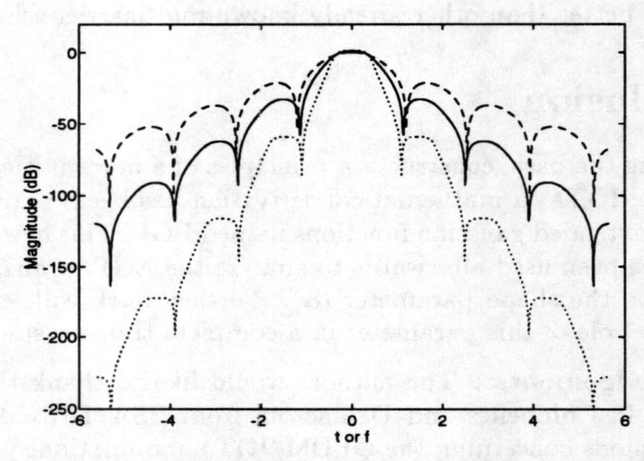

Figure 1: Time and frequency representation of z_α for $\alpha = 1/2$ (dashed line, (in time)), 1 (solid line) and 2 (dotted line, (in time)).

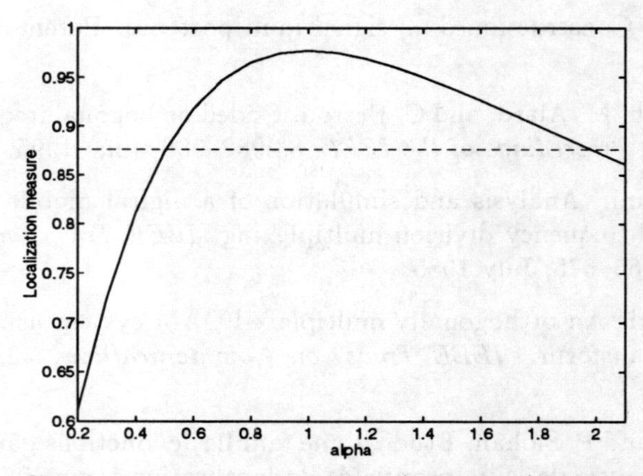

Figure 2: Localization measures ξ for z_α (solid line) and OFDM/OQAM, OFDM/MSK multicarrier schemes (dashed line).

ON THE DUALITY OF MULTI-CARRIER SPREAD SPECTRUM AND SINGLE-CARRIER TRANSMISSION

Karsten Brüninghaus, Hermann Rohling

Technical University of Braunschweig
Schleinitzstraße 22, D-38092 Braunschweig, Germany
E-Mail: rohling@ifn.ing.tu-bs.de

INTRODUCTION

The combination of orthogonal frequency division multiplexing (OFDM) and code division multiple access (CDMA) has found great interest during the last few years. Several ways of combination have been discussed in the literature, named as Multitone-CDMA (MT-CDMA) [1], Multi-Carrier DS-CDMA [2] and Multi-Carrier or OFDM-CDMA [3, 6]

In this paper we will focus on Multi-Carrier CDMA (MC-CDMA). In contrast to a conventional OFDM system where every bit is assigned to a single subcarrier, the information is spread over several subcarriers using orthogonal codes. Since spreading is performed on information symbol level prior to subcarrier modulation, the resulting MC-CDMA signal is basically an OFDM signal. The advantage of this technique is a lower bit error rate in frequency-selective radio channels compared to a conventional OFDM system due to inherent frequency-diversity. The price to be paid is a higher computation complexity in the receiver since the frequency-selective radio channel leads to a loss of orthogonality of the spreading codes and interuser-interferences occur. A lot of research has been done to develop suitable detection or joint detection methods [4, 5, 6, 7] without considering the spreading matrix in detail.

K. Fazel and G.P. Fettweis (eds.), Multi-Carrier Spread-Spectrum, 187-194.

In this paper we will discuss this item focussing on the special case that the spreading matrix equals the Fourier matrix. The two operations performed in the transmitter, namely spreading and IFFT-processing, cancel each other resulting in a pure single-carrier system with blockwise processing. This special case was also mentioned in [8] but little was said about the consequences, which will be done in the following sections.

PRINCIPLE OF MULTI-CARRIER CDMA

In this section we will first give a brief review on the principle of MC-CDMA. Note that although the term "multiple access" occurs in the word MC-CDMA this technique can also be used as a pure transmission technique if all available codes are assigned to a single user. Since the analytical description is identical in both cases we do not make a distinction here.

The general block diagram of a MC-CDMA transmitter (only the downlink is considered in this paper) is depicted in Fig. 1. In a first step the complex data symbols x_i, $1 \leq i \leq M$, which can be elements of any arbitrary symbol set e.g. PSK or QAM, are spread over several subcarriers using orthogonal codes. The spreading can be described analytically by a matrix operation

$$\mathbf{s} = [\mathbf{H}] \cdot \mathbf{x} \tag{1}$$

where $\mathbf{x} = (x_1, \ldots, x_M)^T$ is the vector of data symbols, $\mathbf{s} = (s_1, \ldots, s_N)^T$ is the vector of modulation symbols and $[\mathbf{H}]$ is the code matrix. The M columns of the code matrix contain different spreading codes for every data symbol. To avoid interuser-interferences in AWGN the spreading codes should be mutually orthogonal. If the OFDM system consists of N subcarriers there are a maximum number of N orthogonal codes and $1 \leq M \leq N$ users can be supported every OFDM symbol. Apart from the described spreading procedure, the transmitter is identical to a conventional OFDM transmitter, i.e. the resulting signal sequence is IFFT-processed, periodically extended (a guard-interval is added) and D/A converted.

In the receiver, the incoming signal is sampled and Fourier-transformed. Since the orthogonality of the spreading codes is destroyed in frequency-selective radio channels and interuser-interferences occur, sophisticated detection algorithms have to be applied. All these detection methods require knowledge of the channel influence and

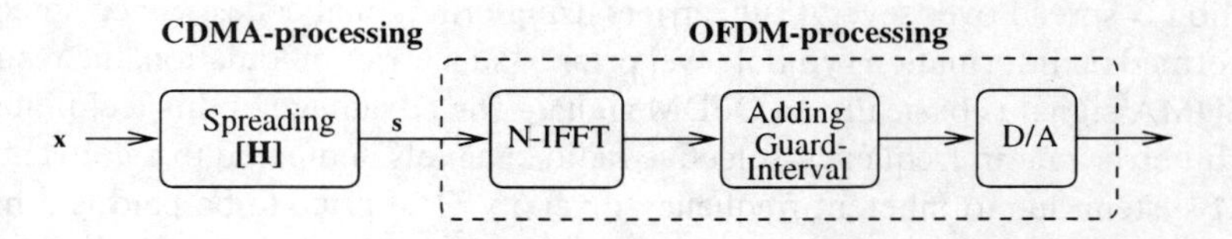

Figure 1: Block diagramm of a MC-CDMA transmitter (downlink)

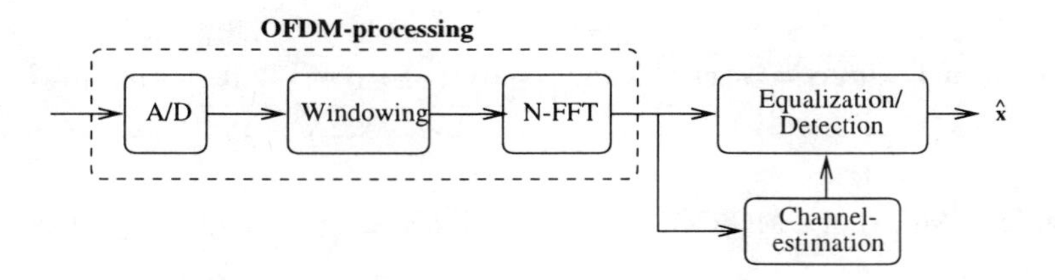

Figure 2: Block diagramm of a MC-CDMA receiver (downlink)

therefore channel estimation has to be performed in the receiver, too. A general block diagram of the receiver is depicted in Fig. 2.

In this paper we assume that a linear equalizer is used in the receiver to reduce the interuser-interferences. With this assumption the overall transmission system is completely linear and can be modelled by a series of matrix multiplications as depicted in Fig. 3. Due to the periodic extension of time signal in the transmitter and windowing in the receiver, the channel influence can be modelled by a cyclic convolution in the time domain or a simple multiplication in the frequency domain. Assuming ideal synchronization in the receiver, the channel transfer matrix $[\mathbf{K}]$ simplifies to a diagonal matrix.

SPREADING CODE SELECTION AND CONSEQUENCES

In most of the analytical work concerning MC-CDMA, orthogonal Hadamard codes were assumed. These codes are optimal in the sense that the energy of every information symbol is uniformly distributed on every subcarrier leading to a minimal mean square error after linear equalization in frequency-selective radio channels. In addition, it was discussed to influence the dynamic range of the time signal by properly selecting the code matrix. For the uplink suitable codes have been found in [9] reducing the dynamic range of the time signal of every user to 6 dB.

Both demands, i.e. minimal mean square error in frequency-selective fading channels and low dynamic range of the signal envelope, can also be met for the downlink if an FFT matrix is used as spreading matrix. In this case spreading and IFFT-processing cancel each other, see Fig. 3, and the transmission system degenerates into a single-carrier system, as depicted in Fig. 4. The main advantage of this step is that at this stage all knowledge about dynamic reduction of single carrier systems with linear modulation can be applied to create a time signal with low dynamic range.

Fig. 5a shows as an example the signal envelope of a MC-CDMA system with Walsh-Hadamard codes and 4-PSK modulation. Using the FFT matrix as spreading matrix reduces the dynamic range of the envelope, see Fig. 5b, but the signal is still far away from having a constant amplitude due to the $\sin(x)/x$ pulse shape. A further reduction of the time signal dynamic can be achieved if a $\pi/4$-QPSK with square-root raised cosine puls spectrum is used as shown in Fig. 5c. Other methods like for

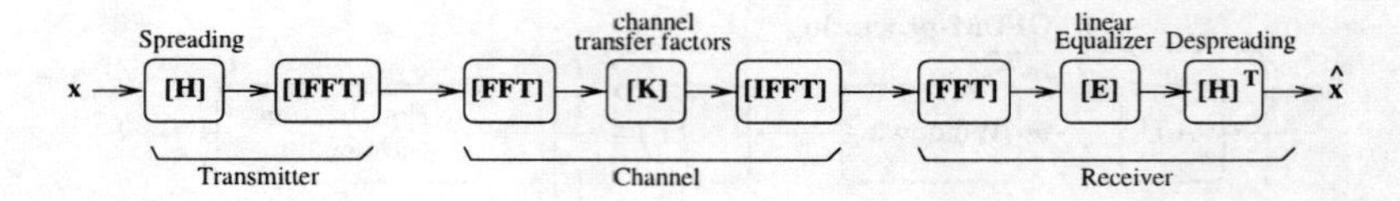

Figure 3: Model of a MC-CDMA transmission system with linear equalization

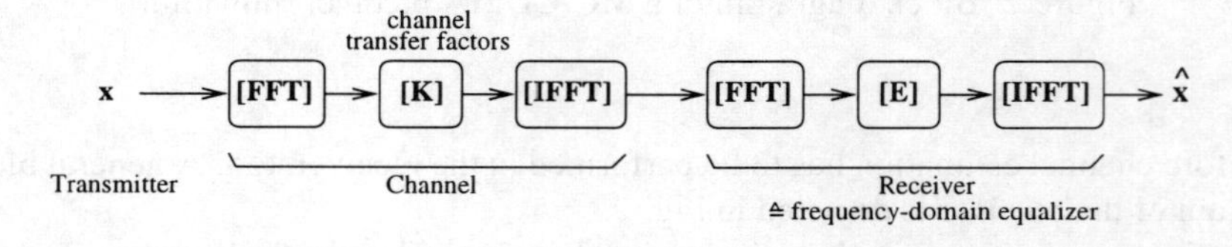

Figure 4: MC-CDMA transmission system with FFT spreading matrix

example the use of offset QPSK or MSK (in case of binary signalling) modulation further reduce the signal dynamic considerably.

Apart from solving the signal dynamic problem, the single-carrier system resulting from MC-CDMA with FFT spreading matrix shows some other advantages which are discussed in the following.

On the one hand, the multi-carrier inherent high sensibility to phase noise and frequency/clock synchronization errors are reduced, allowing simpler synchronization circuits.

On the other hand, the single-carrier system is still a special case of a MC-CDMA system and therefore the same performance is to be expected. Using a linear equalizer for MC-CDMA as discussed here results in a frequency domain equalizer for the corresponding single-carrier system. Such a frequency domain equalizer was originally proposed in [10] having in mind a faster convergence of the equalizer taps if the stochastic gradient algorithm is applied in the frequency domain. Recently this fre-

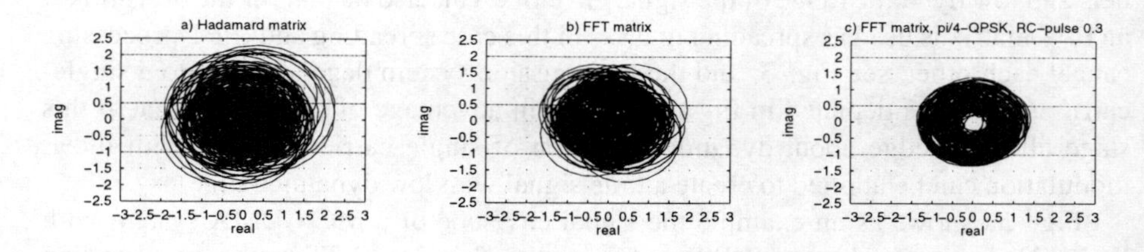

Figure 5: Envelope dynamics for a MC-CDMA system with a) Hadamard spreading matrix b) FFT spreading matrix and c) FFT spreading matrix with $\pi/4$ QPSK and RC-pulse spectrum

quency domain equalization was picked up again in [11] to compare single- and multi-carrier transmission systems but without noticing that the proposed single-carrier system is also a special form of a MC-CDMA system.

In contrast to a conventional single-carrier system, the single carrier system resulting from MC-CDMA uses a guard interval. As already mentioned this guard interval (and the corresponding signal windowing in the receiver) causes a cyclic convolution of the channel impulse response and the transmitted signal. If the corresponding discrete-time channel transfer function contains no zeros, this cyclic convolution can be perfectly inverted in the receiver. To achieve the same effect with a conventional single carrier system an equalizer with infinite taps is required which cannot be implemented in a real system. Therefore, the single-carrier system with guard-interval outperforms the comparable single carrier system without guard interval for the same number of equalizer taps which is further analysed in the following. The minimum mean square error for any linear equalizer is given by [12]

$$J_{\min} = 1 - \varepsilon^{T*}[\Gamma]^{-1}\varepsilon \tag{2}$$

where

$$[\Gamma] = E\{(\mathbf{v}^*\mathbf{v}^T)\} \qquad \text{and} \qquad \varepsilon = E\{\mathbf{v}^* I\} \tag{3}$$

Here, $\mathbf{v}$ is the input sequence and I the output value of the linear equalizer. In case of a single-carrier system without guard interval, the covariance matrix $[\Gamma]$ is Hermitian having the general shape depicted in Fig. 6a. For the single carrier system with guard interval, every line of the covariance matrix is additionally periodic modulo the matrix dimension leading to the shape shown in in Fig. 6b.

An analysis of $J_{\min}$ or equivalently the signal-to-noise ratio (S/N) at the output of the linear equalizer is performed for the generalized symmetric discrete time channel $f_1, f_0 = f_2 = \sqrt{(1-f_1^2)/2}$. The input noise is set to zero. Especially if the channel has deep spectral fades - which is the case for $f_1 < 0.8$ - the single carrier system without guard interval fails to invert the channel transfer function resulting in an irreducible error floor, see Fig. 7 right. An increase of the equalizer taps only slightly reduces the irreducible error flow as shown in Fig. 7 left, for the fixed channel $f_1 = 0.8, f_0 = f_2 = 0.424$. $J_{\min}$ is zero, i.e. no error floor occurs, if a single carrier system with guard interval is used.

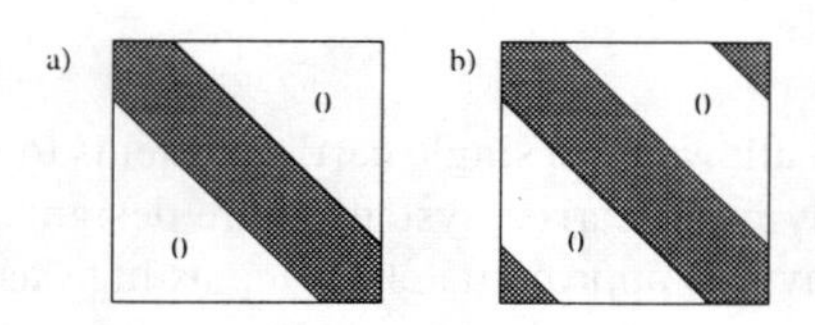

Figure 6: Shape of the covariance matrix Γ for single carrier systems a) without and b) with cyclic extension

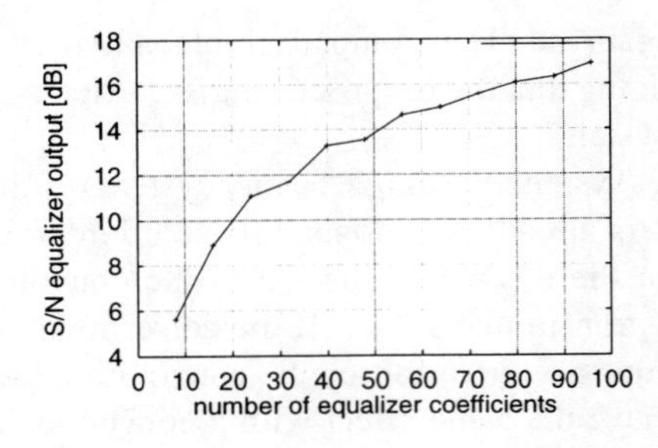

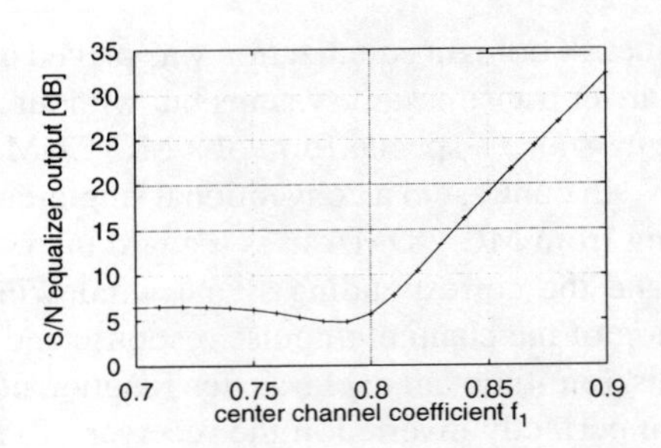

Figure 7: S/N at the output of a conventional single-carrier system with linear equalization as a function of the number of taps (left figure, $f_1 = 0.8, f_0 = f_2 = 0.424$) and as a function of the channel impulse response (right figure, 8 taps); no input noise

This behaviour was also measured in simulations using a stochastic channel model, defined by an exponentially decreasing power delay profile, see Fig. 8. The performance loss of the single-carrier system with guard interval and linear equalization compared to an AWGN channel amounts to only 6.5 dB (at BER=10^{-4}). If antenna diversity is used in the receiver the loss can be further reduced to 3.5 dB.

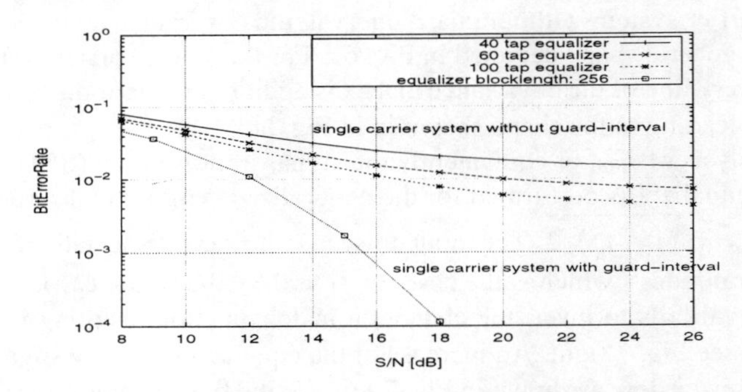

Figure 8: Bit error curves of a single-carrier system with and without guard interval using a linear equalizer (MMSE-criterion); frequency-selective radio channel, time-invariant, blocklength $T_N = 35\mu s$,

Adding a guard interval allows even single carrier systems to cope with very long impulse responses originally multi-carrier systems were designed for. The computation complexity in the receiver is approximately twice as high as using conventional coherent OFDM which has - at least in the uncoded case - a worse performance.

CONCLUSIONS

The special choice of a FFT matrix as spreading matrix for Multi-Carrier CDMA leads to a single-carrier system with guard-interval. This way, the multi-carrier specific problems as high dynamic of the signal envelope, sensivity to phase noise and frequency/clock synchronization errors can be avoided without losing the performance capacity of MC-CDMA. In case of linear equalization the receiver consists of a frequency domain equalizer with blockwise processing. Due to the guard-interval, this frequency domain equalizer outperforms the comparable time domain equalizer considerably and shows even in frequency-selective radio channels a relatively good performance. Besides, the compuation complexity of such a receiver structure with blockwise processing in the frequency domain is lower than with processing in the time domain in case of highly dispersive channels.

References

[1] L. Vandendorpe: "Multitone Spread Spectrum Multiple Access Communications System in a Multipath Rician Fading Channel". IEEE Trans. on Veh. Tech., Vol.44, No.2, May 1995

[2] V.M. DaSilva, E.S. Sousa: " ".Proc. of IEEE ICUPC'93, pp. 995-999, Ottawa, Canada, Oct. 1993

[3] N. Yee, J.P. Linnartz, G. Fettweis: "Multi-Carrier CDMA in Indoor Wireless Radio Networks". Proceedings PIMRC '93, Yokohama, Japan, 1993, pp. 109-113

[4] N. Yee, J.P. Linnartz: Controlled Equalization of Multi-Carrier CDMA in an Indoor Ricean Fading Channel. Proceedings PIMRC' 94, Stockholm, 1994, pp. 1665-1669

[5] N. Yee, J.P. Linnartz: Wiener Filtering of Multi-Carrier CDMA in Rayleigh Fading Channel. Proceedings PIMRC '94, Stockholm, 1994, pp. 1344-1347

[6] T. Müller, H. Rohling, R. Grünheid: Comparison of different Detection Algorithms for OFDM-CDMA in Broadband Rayleigh Fading. Proceedings IEEE VTC '95, Chicago, USA, pp. 835-838

[7] M. Reinhardt, J. Hess, J. Lindner: Combined Decision Feedback Equalization and TCM-Decoding for Multiple Input/Multiple Output Channels with Applications in CDMA-Systems. Proc. Third UK/Australian International Symposium on DSP for Communication Systems, University of Warwick, UK, 12-14 Dec. 1994

[8] J. Lindner: MC-CDMA and its Relation to General Multiuser/Multisubchannel Transmission Systems. Proc ISSSTA'96, Mainz, 1996, pp. 115-121,

[9] V. Aue, G.P. Fettweis: "Higher-Level Multi-Carrier Modulation and its Implementation". Proc ISSSTA'96, Mainz, 1996, pp. 126-130,

[10] T. Walzman, M. Schwartz: "Automatic equalization using the discrete Fourier domain". IEEE Trans. Inform. Theory, vol.IT-19, pp.59-68, January 1973

[11] H. Sari, G. Karam, I. Jeanclaude: An Analysis of orthogonal frequency division multiplexing for mobile radio applications. 1994 IEEE.

[12] J.G. Proakis: "Digital Communications", sec. Ed. McGraw-Hill series in Electrical Engineering.

MISMATCHED MULTI CARRIER COMPLEMENTARY SPREAD SPECTRUM RADAR AND SONAR SYSTEMS

Igor S. Simić[1], Aleksa J. Zejak[1], Miroslav L. Dukić[2] and Miljko M. Erić[3]

[1]*IMTEL - Institute of Microwave Techniques and Electronics, B. Lenjina 165b, 11 070 Belgrade, Yugoslavia; E-mail: zejak@imtel.co.yu*
[2]*Department of Communications, Faculty of Electrical Engineering, B. Revolucije 73, 11000 Belgrade, Yugoslavia;*
[3]*Institute of electrical engineering VTI, Katanićeva 15, Belgrade, Yugoslavia*

ABSTRACT

In this paper MC-SS radar or sonar concept that make it possible complementary sequences application is proposed. An ideal autocorrelation properties of this sequences are degraded for greater target speeds. To improve autocorrelation ambiguity function DIRLS algorithm for filter response shaping is applied.

I. INTRODUCTION

Complementary sequences have the property that when their individual range sidelobes are combined, the composite sidelobes are completely cancelled, yielding the desired perfect correlation property. A lot of papers [1,2,3] deal with a construction of complementary sequences. However, for non-zero Doppler shifts

K. Fazel and G.P. Fettweis (eds.), Multi-Carrier Spread-Spectrum, 195-200.

there are sidelobes of the combined auto-correlation function that make complementary sequences radar application senseless. Hence, the most of authors who deal with radar complementary sequences application suggest their indirect use naming them: "subcomplementary", "supercomplementary", "group-complementary" etc. [4,5,6].

In this paper we proposed complementary sequences application concept of Multi-Carrier Spread Spectrum (MC-SS). We described this radar concept and analysed the main characteristics of the signal. For sidelobe suppression in Doppler range of ambiguity function we suggest Doppler optimised mismatched filter (MMF). We achieved interesting results.

II. MC-SS RADAR CONCEPT

A radar or sonar MC transmitter spreads the original pulse over different subcarriers using given complementary spreading codes (Fig. 1). The number N of spreading complementary codes is equal to the number of subcarriers.

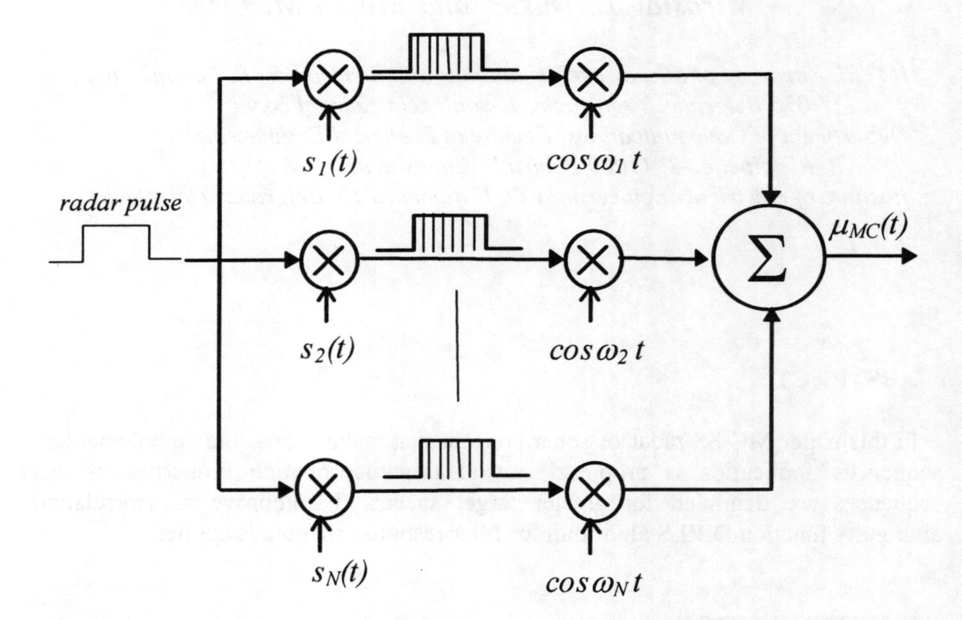

Fig. 1 MC radar or sonar transmitter scheme

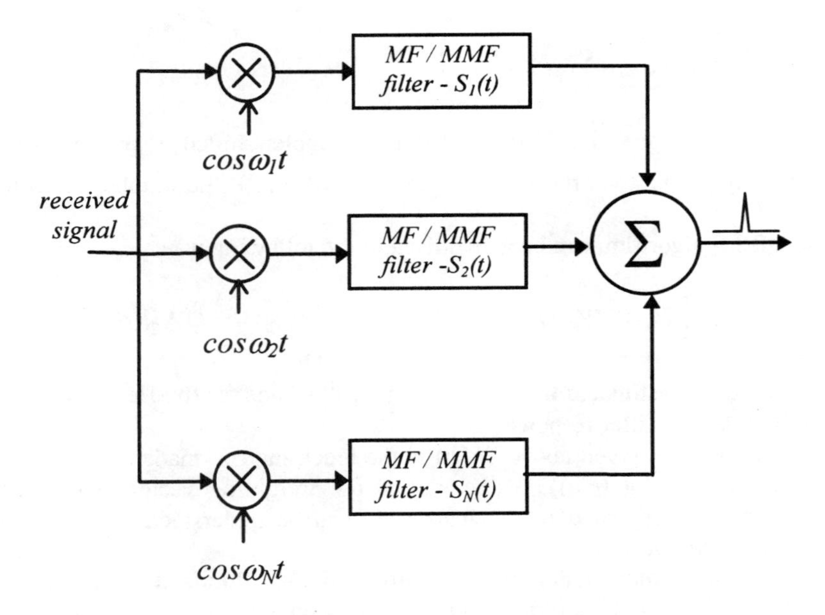

Fig. 2 MC radar or sonar receiver scheme

Fig. 2 shows a MC radar or sonar receiver, composed of N coherent receivers, each one matched to the one of the par or set of complementary sequences. For the zero Doppler shift the sequences are passed through their individual matched filters (MF) and then added, which results in the cancellation of sidelobes and the addition of the mainlobes.

However, for greater target velocities, sidelobes are not cancelled. The sidelobe level increases very fast by the target velocity increase. In order to cancel sidelobes after summing up individual sequences, appropriate mismatched filters are designed. Each individual response of the mismatched filter for extended target velocity range should be as close as possible to the matched filter response for the zero Doppler shift. Mismatched filters are designed by using the Doppler optimised IRLS algorithm.

III. DOPPLER OPTIMIZED IRLS ALGORITHM

The optimisation procedure for the mismatched filter in a given Doppler band can be defined as a procedure of forming a suitable shape of the ambiguity function. Unlike standard filters, where the object of shaping is the correlation function, in this type of filters the object is the ambiguity function, or rather its sector.

Also, the block matrix corresponds to the signal matrix,

$$S_\Phi = \left(S_{f_1}, \ldots, S_{f_i}, \ldots, S_{f_P}\right)^T, \tag{1}$$

where S_{f_i} is the signal matrix for a particular Doppler shifted frequency, N is the sequence length and M is the filter length ($M \geq N$) and P the number of Doppler shifts.

The DIRLS algorithm can be generalised in the following way:

$$\hat{x}(n) = \left[S_\Phi^H(0) W_\Phi(n-1) S_\Phi(0)\right]^{-1} S_\Phi^H(0) W_\Phi(n-1) \Delta_\Phi(n-1) \tag{2}$$

where $\hat{x}$ stands for estimated filter coefficients, $[.]^H$ stands for the Hermitian matrix and Δ_Φ for a desired filter response.

In the expression given above W (n) is the block matrix, made of the diagonal matrices R (n) = diag ($r(n)$), where $r(n)$ is the weighting vector. The window function, which is included in the matrix W, can be understood as a corrective factor of the LS algorithm.

The analysis of implementation of the proposed algorithm is also carried out for the most significant and well known phase coded signals in radar systems, such as Frank polyphase sequences, P1, P2, P3, P4, Barker's binary sequences and others.

In [11] we considered a possibility of sidelobe suppression, i.e. self clutter suppression in modern radar and sonar systems with intrapulse frequency hopping. We adapted the Doppler optimised IRLS algorithm in order to shape the ambiguity function of frequency hopping sequence based upon Costas arrays.

IV. DIRLS APPLICATION IN MC-SS RADARS OR SONARS - RESULTS

The DIRLS (2) algorithm is applied to mismatched filter response shaping. The basic difference in applying this algorithm to the sidelobe suppression of DS and FH sequences is the value of Δ_Φ desired filter response. In the case of complementary sequences, the sidelobe suppression in each of the MC receiver's branches after the summing gives worse results than the DIRLS algorithm application to mismatched filter response shaping, being as close as possible to the matched filter response for the zero Doppler shift. The value of Δ_Φ desired filter response is instead Dirac pulse equal to matched filter response for zero Doppler shift.

For the Golay complementary pair of sequences (N=2; $\{s_1\}$={1,-1,1,1,-1,1,1,1}; $\{s_2\}$={1,-1,1,1,1,-1,-1,-1}) of the length 8, we analyse the proposed method for improvement of combined ambiguity function properties at the Doppler range.

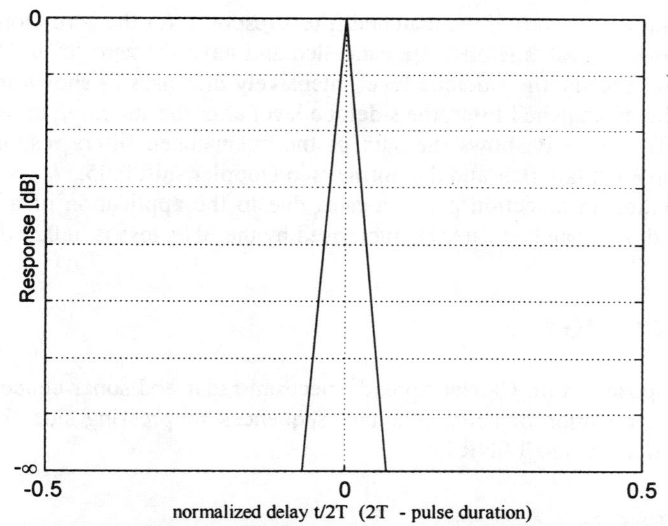

Fig. 3 Combined autocorrelation function - the sum of the matched filters response for the Golay complementary pair and the zero Doppler shift

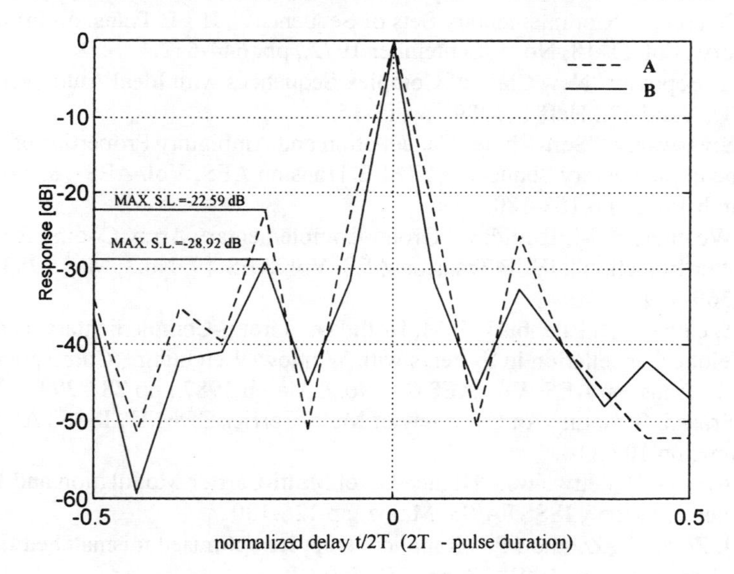

Fig. 4 Sum of: **A** - the matched filters responses, **B** - the mismatched DIRLS filters responses for the Golay complementary pair and 0.05 Doppler shift normalised to the 1/T (T-pulse duration)

Fig. 3 shows the sum of the matched filters response for the zero Doppler shift. It can be shown that all sidelobes are cancelled and have the zero level. However, for higher target speeds, the sidelobe level intensively increases as shown in the Fig. 4-A. Using the mismatched filter, the sidelobe level after the summing up is suppressed for 6.33 dB. Fig. 4-B shows the sum of the mismatched filters responses for the Golay complementary pair and the normalised Doppler shift 0.05.

Degradation in detection performances due to the application of a mismatched filter instead of a matched filter characterised by the SNR loss is 0.025 dB.

V. CONCLUSION

The proposed Multi Carrier Spread Spectrum radar and sonar concept enables a successful application of complementary sequences suppressing sidelobes for great Doppler shifts and small SNR loss.

REFERENCES:

[1] M.J.E. Golay: "Complementary series", IRE Trans., 1961, IT-11, pp. 207-214

[2] C.C. Tseng : "Complementary Sets of Sequences", IEEE Trans. on Information theory, Vol. IT-18, No. 5, September 1972., pp. 644-652.

[3] B.M. Popović: "New Class of Complex Sequences with Ideal Autocorrelation", AEU, Band 43., Heft 1.,1989., pp.13-15.

[4] R. Sivaswamy: "Self-Clutter Cancellation and Ambiguity Properties of Subcomplementary Sequences", IEEE Trans.on AES, Vol. AES-18, No.2, March 1982., pp.163-180.

[5] G. Weathers, E.M. Holliday: "Group-Complementary Array Coding for Radar Clutter Rejection", IEEE Trans.on AES, Vol. AES-19, No.3, May 1983., pp.369-379.

[6] G. Weathers, D. Ellenburg, E.M. Holliday: "Group-Complementary Temporal Sidelobe Cancellation in Systems with Window-Weighting", Correspondence, IEEE Trans. on AES, Vol. AES-23, No.2, March 1987., pp.287-293.

[7] R. Prasad, S. Hara: "An Overview of Multi-Carrier CDMA", ISSSTA'96 Mainz, pp.107-114.

[8] V. Aue, G. P. Fettweiws: "Higher-Level Multi-Carrier Modulation and Its Implementation", ISSSTA'96, Mainz, pp.126-130.

[9] A. J. Zejak, E. Zentner, P.B.Rapajić: "Doppler optimised mismatched filters", Electronics letters, 1991, 27, pp. 558-560

[10] P. B. Rapajić, A. J. Zejak: "Low sidelobe multilevel sequences by minimax filter", Electronic letters., 1989, 25, pp. 1090-1091

[11] A. J. Zejak, I.S. Simić, M.L. Dukić: "Frequency Hopping Mismatched Filters for Radar and Sonar Applications", ISSSTA'96 Mainz, pp.877-881.

Section V

MC-SS FOR THE UPLINK OF A CELLULAR SYSTEM

AN OFDM/SFH-CDMA TRANSMISSION SCHEME FOR THE UPLINK[1]

*Luciano Tomba** *Witold A. Krzymien***

*University of Padova (Italy) **University of Alberta/TR*Labs*[†]

[†]800 Park Plaza, 10611 - 98 Avenue, Edmonton (AB), T5K 2P7 CANADA

e-mail: wak@edm.trlabs.ca

Abstract

The demand for portable/mobile transmission equipments for multimedia applications calls for the development of new schemes in principle capable of combining the benefits of Time Division Multiple Access (TDMA) and Code Division Multiple Access (CDMA). Multicarrier-CDMA (MC-CDMA) seems to be an interesting candidate because it is able to ensure good performance in fading channels at reasonable receiver complexity and, at the same time, it allows high rate transmission. While the design of similar modem schemes for the *downlink* has been widely investigated (and many results have confirmed their feasibility), fewer results are available for the *uplink*. In this contribution we analyze some proposals available in the literature and then we present another possible scheme for the uplink.

1 INTRODUCTION

For the last several years the use of wireless communication systems has been growing at an astonishing rate. After the definition of well known second generation standards like GSM (in Europe) and IS-95 (in the USA) for cellular telephony, the interest in high rate transmission schemes for multimedia applications in micro-cells (or even larger cells) is now emerging (e.g. interactive television systems for rural areas), and the use of Orthogonal Frequency Division Multiplexing (OFDM) is seen as a reasonable alternative solution to overcome the problems caused by the frequency selectivity of the radio channel [1]. In this paper we confine our attention to the uplink. First we present the most significant proposals for the uplink available in the recent literature; then, we present our proposal which is based on the combination of CDMA obtained by slow frequency hopping (SFH) and OFDM. OFDM is introduced as a means of mitigating multipath effects of the radio channel, mainly in outdoor residential environments [2, Ch. 3]. Performance of the proposed scheme is discussed in the presence of antenna space diversity, block coding and packet combining. The motivating factor behind the proposed scheme, and SFH in particular, is the very stringent synchronization requirement of direct sequence CDMA systems. In fact, simulation results show that the performance of a MC-CDMA scheme as proposed in [3] and later analyzed in [4] is very poor when applied to the uplink where, because of its asynchronous character, significant interference arises as the number of users increases. In order to avoid complex successive interference cancellation schemes, here we propose the combination of SFH and OFDM together with some techniques useful to improve performance. The purpose of SFH is to introduce multiple access capability to the

[1]Funding for this work was provided, in part, by the Telecommunications Research Laboratories (TR*Labs*), the Natural Sciences and Engineering Research Council of Canada and the "Consiglio Nazionale delle Ricerche," Roma (Italy).

K. Fazel and G.P. Fettweis (eds.), Multi-Carrier Spread-Spectrum, 203-210.

system, while OFDM makes the system robust against multipath fading. We note that a system like this one is reasonable only if very high data rate must be assured, e.g. in multimedia applications. At data rates of a few Mbit/s or less, in the authors' opinion the MC-CDMA technique is not justified and more conventional schemes such as TDMA or DS-CDMA should be used.

2 OVERVIEW OF EARLIER UPLINK PROPOSALS

2.1 Kondo-Milstein's Scheme

In the Kondo-Milstein's scheme [5] a data sequence is multiplied by a spreading sequence and it modulates M subcarriers rather than one single carrier. At the receiver end, each subcarrier is demodulated and correlated with the pseudonoise (PN) sequence, and the outputs of the correlators are combined using the maximal ratio combining (MRC) technique. The advantage of this solution with respect to the single carrier direct sequence-spread spectrum (DS-SS) scheme is that each subcarrier is modulated by a spreading sequence with a chip duration M times longer than that of a single-carrier system. On the other hand, MRC requires estimation of the gain of each path, the system complexity linearly increases with M, and it seems to be impractical to consider $M > 10$.

2.2 Sourour-Nakagawa's Scheme

The Sourour-Nakagawa's scheme [6] is not much different from the one discussed above, because it can still be put in the class of systems which use a parallel transmission of narroband DS waveforms in the frequency domain. However, in this scheme first the source data stream is serial-to-parallel converted to a number of lower rate streams; then each stream feeds a number of parallel streams with the same rate. The bits of these streams are then interleaved (to get advantage from time diversity) and spread by the same PN code. However, also this scheme seems to be overly complex.

2.3 Chen-Sousa-Pasupathy's Scheme

The scheme proposed by Chen, Sousa and Pasupathy [7] is similar to the one proposed in [5]. However, it no longer uses frequency diversity transmission (equivalent to a rate $\frac{1}{M}$ repetition code) which may reduce the spectral efficiency of the system. The idea presented in [7] is to design an algorithm capable of choosing for each substream the best subchannel with respect to the fading process, and hence introducing a form of selection diversity. However, since more than one substream can choose a subchannel with a better fading margin, the substreams are spread by a PN sequence prior to the selection in order to ensure unique demodulation. Apart from the difficulty of implementing an efficient algorithm to select the best subchannels, the advantage of simultaneously transmitting a high number of substreams on the same subchannel is not clear. Moreover, an OFDM system with non-uniform load among subcarriers seems to be more sensitive to possible system imperfections. In particular, it would be interesting to compare performance of the three systems described in this Section under imperfect timing and frequency synchronization.

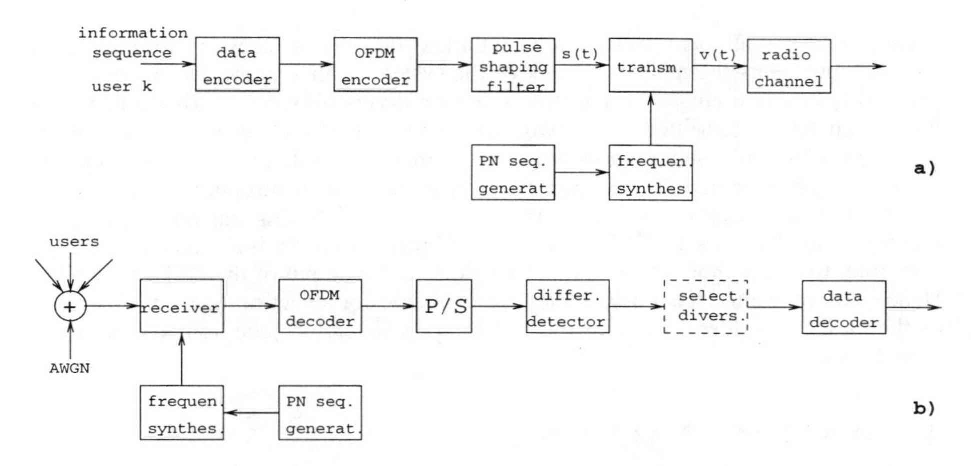

Figure 1: Block diagram of the proposed system.

3 THE PROPOSED SYSTEM

3.1 Transmitter

The transmitter block diagram is shown in Fig. 1a. The main features of the proposed system are:
i) Use of OFDM which, in principle, allows high rate transmission without complex equalization.
ii) Use of SFH allowing multiple access capability of the system with potentially better performance than that of DS-CDMA because of the stringent synchronization requirements of the latter.
iii) The hopping time equal to the sum of the duration of one OFDM symbol and one guard interval, in order to avoid disturbing differential detection performed within each OFDM symbol by frequency hops occurring within the OFDM symbol intervals.
iv) Frequency diversity introduced by frequency hopping, which helps to overcome fading.

We consider a hypothetical system with a carrier frequency at around 40 GHz. The available bandwidth is divided into 32 frequency slots and each slot can support a bit rate of 10 Mbit/s; hence, with DQPSK modulation the symbol period is 200 ns. If we assume a channel with a total excess delay of up to 1000 ns, a guard interval of length $G = 5T$ should be sufficient to avoid intersymbol interference. With a DFT size of 64, the loss in spectral efficiency due to the guard interval is limited to 7.8%.

3.2 Receiver

In the reverse link the receiver is at the base station. Hence, the receiver complexity is not as important an issue as in the forward link, where the receiver is in the user's unit. The proposed scheme requires a multiuser receiver with 32 receiving filters corresponding to the 32 different frequency slots. To demodulate the signals corresponding to all users, in principle an OFDM decoder is necessary for each user, and this represents the main drawback of the system. Our scheme also includes the possibility of employing post-detection selection diversity (see Fig. 1b, dashed box) which can be implemented using two different antennas and receivers. The post-detection selection scheme works much better than a hypothetical pre-detection one because in the latter case the selection would be performed on the composite frequency domain representation of the transmitted signal. To clarify this statement, we recall that the equivalence of the OFDM system to N parallel flat fading channels holds, in principle, from the input of the OFDM encoder to the output of the OFDM decoder. Hence, the relevant decision variables are available for selection only at the output of the OFDM decoder. The drawback of the post-detection scheme is its hardware complexity.

4 SYSTEM ANALYSIS

Our system is equivalent to an OFDM modem within one single hopping period. Hence, firstly we analyze the OFDM scheme and secondly we present the new features introduced by the SFH technique.

4.1 Analysis of the OFDM Scheme

The OFDM system discussed here transmits differentially encoded data; the demodulation is possible using the information transmitted on two adjacent subcarriers. The first subcarrier is used to transmit a reference sequence. For $0 \leq t < NT$, the transmitted signal is $v(t) = s(t)\, e^{j(2\pi f_0 t + \theta_0)}$ where f_0 and θ_0 are the frequency and the phase of the carrier corresponding to the zeroth DFT block. A guard interval of G symbol intervals is inserted before the next hopping period. The received signal is correlated with tones of the same frequency as in the transmitter. Consequently, the samples at the output of the OFDM decoder are

$$z_i = d_i \lambda_0\, C_i + \sum_{\substack{n=0 \\ n \neq i}}^{N-1} d_n \lambda_{n-i}\, C_n + \eta_i \quad , \qquad 0 \leq i < N \quad , \tag{1}$$

where C_n is the nth channel gain,

$$\lambda_n \triangleq \frac{1}{NT} \int_{\max\{0,\varepsilon\}}^{\min\{NT, NT+\varepsilon\}} e^{j 2\pi n \frac{t}{NT}}\, dt \quad , \tag{2}$$

η_i are samples of the additive white gaussian noise (AWGN), and ε is an independent random variable modelling the error in the hopping instant. The second term in (1) represents the inter-subchannel interference (ICI) due to a lack of synchronization represented by the error term ε. When $\varepsilon = 0$, $\lambda_n = 0$ for $n > 0$ (see (2)), and the ICI term disappears[2].

[2]This is true if no other impairments exist (e.g. oscillator frequency offset and phase noise, or guard interval shorter than the channel impulse response).

4.2 Performance Improvement Strategies

Antenna Diversity Selection Scheme. To improve the performance of the OFDM scheme, we adopt a post-detection antenna diversity selection scheme and we assume that the signals from the two antennas (labelled A and B) are uncorrelated. Our selection strategy is the following: let y_i^A and y_i^B be the two decision variables corresponding to antenna A and antenna B, respectively; then the selected decision variable is $y_i = \max\{|y_i^A|, |y_i^B|\}$. It is apparent that this decision rule is sub-optimum.

Block Encoding. The proposed antenna diversity scheme is effective to mitigate fading, but it is not able to remove the correlation of errors between the subcarriers. Here we propose the combination of coding, sub-carrier interleaving and antenna space diversity to reduce the error rate. We assume that 64 bits of information are coded using a block code (in particular, a linear $(8, 4)$ binary block code is considered). After FEC coding the 128 code bits are DQPSK modulated and interleaved with interleaving depth of 8 before the OFDM encoder. The interleaving of the data modulating the subcarriers is useful to reduce the correlation of errors on adjacent subcarriers. Additionally, to reduce the error probability due to the interference produced by other users simultaneously transmitting on the same frequency slot, we may use an ARQ procedure and packet combining.

Packet Combining. In frequency hopping spread spectrum the number of errors introduced into a packet depends on the number of simultaneously active users. However, a particular feature of this multiple access technique is that only a portion of the packet is affected by errors if the system is asynchronous [8]. Hence, the throughput of the conventional system can be improved by combining the different received copies of the same packet. Moreover, this strategy is effective against fading because, due to the large bandwidth of the system, it is not unreasonable to consider the fading process on different frequency slots to be uncorrelated.

5 RESULTS

In this section we discuss some results which we have obtained for the proposed system. The channel model considered is the worst case *indoor commercial* equivalent to the best case *outdoor urban/suburban* in accordance with the classification in [2, Table 6B.2, 6B.8]. We have assumed all the paths' amplitude to be Rayleigh distributed and we have presented performance in terms of bit and packet error probability.

Fig. 2 shows the bit error rate (BER) versus the index of the sub-carriers. The gain obtained through diversity is apparent. However, even with diversity there is significant correlation between the BER values of adjacent sub-carriers. The figure shows results for just one static channel, but the conclusion is the same for different channel realizations. This behaviour implies that interleaving within the same OFDM block and error correction coding should reduce the BER and ensure a more flat shape of the curves versus the sub-carrier index. Initially we assume that only one user is active; hence, no multiple access interference exists. Fig. 3 shows the average BER (i.e. the mean value of the BER over all sub-carriers) versus the SNR/bit. Different configurations of the receiver are considered. In particular, without the use of diversity and coding the performance is quite poor. Some enhancement in system performance is achieved by the use of coding, and the use of interleaving within the same OFDM block brings further improvement. However, the most significant gain is represented by the use of antenna diversity which also helps to get a higher gain from the use of coding combined with interleaving. At SNR/bit = 25 dB, use of antenna diversity combined with coding and interleaving gives an improvement of over 3 orders of magnitude in the BER. Finally, Fig. 4 shows the packet error probability for different system configurations, including antenna diversity and packet combining. If packet combining is used, the re-transmitted packet is combined with the previously received packets to take advantage of the intrinsic frequency diversity of the system. The results have been obtained for the case when combining takes place after only one re-transmission (only two packets are combined, and both contain detected errors). It is obvious that the combination of more packets should further enhance performance at the expense of a larger memory capacity. It is clear that the joint use of antenna diversity and packet combining significantly improves performance.

6 CONCLUSION

In this paper we have considered the combination of OFDM and SFH-CDMA for high data rate transmission in microcellular environments. In particular, we have addressed the issue of the design of an efficient scheme for the uplink. The main results can be summarized as follows. *i)* Inter-subcarrier differential modulation seems appropriate because it avoids the problem of frequency domain channel estimation and is more robust against the oscillator phase noise and frequency offset. *ii)* Our investigation has also shown that performance of the OFDM scheme is strongly dependent on the subcarrier index. This known problem has been mitigated by the joint use of coding, subcarrier interleaving and antenna diversity. Further gain has been achieved by the use of packet combining in the ARQ scheme. *iii)* The choice of the hopping frequency patterns plays a fundamental role in the system performance and further research on the subject is required.

References

[1] H. Sari, G. Karam, and I. Jeanclaude, "Transmission techniques for digital terrestrial TV broadcasting," *IEEE Commun. Magazine*, vol. 33, no. 2, pp. 100–109, Feb. 1995.

[2] K. Pahlavan and A.H. Levesque, *Wireless Information Networks.* New York, J. Wiley & Sons, 1995.

[3] N. Yee, J.P. Linnartz and G. Fettweis, "Multi-carrier CDMA in indoor wireless radio networks," in Proc. *PIMRC '93*, pp. 109–113, Yokohama, Sept. 1993.

[4] L. Tomba, W.A. Krzymien, "Downlink detection schemes for MC-CDMA systems in indoor environments," *IEICE Trans. Commun.*, vol. E79-B, no. 9, pp. 1351-1360, Sept. 1996.

[5] S. Kondo and L.B. Milstein, "Performance of multicarrier DS CDMA systems," *IEEE Trans. on Commun.*, vol. 44, no. 2, pp. 238–246, Feb. 1996.

[6] E.A. Sourour and M. Nakagawa, "Performance of orthogonal multicarrier CDMA in a multipath fading channel," *IEEE Trans. on Commun.*, vol. 44, no. 3, pp. 356–367, Mar. 1996.

[7] Q. Chen, E.S. Sousa, S. Pasupathy, "Multicarrier CDMA with adaptive frequency hopping for mobile radio systems," *IEEE Journ. on Sel. Areas in Commun.*, vol. 14, no. 9, pp. 1852–1858, Dec. 1996.

[8] A.M.Y. Bigloo, T.A. Gulliver, and V.K. Bhargava, "A slotted frequency-hopped multiple-access network with packet combining," *IEEE Journ. on Sel. Areas in Commun.*, vol. 14, no. 9, pp. 1859–1865, Dec. 1996.

[9] J.G. Proakis, *Digital Communications*, 3rd ed., New York, McGraw-Hill, 1995.

[10] C.A.F. Wijffels, H.S. Misser, R. Prasad, "A micro-cellular CDMA system over slow and fast Rician fading radio channels with forward error correction coding and diversity," *IEEE Trans. Vehicul. Technol.*, vol. 42, no. 4, pp. 570–580, Nov. 1993.

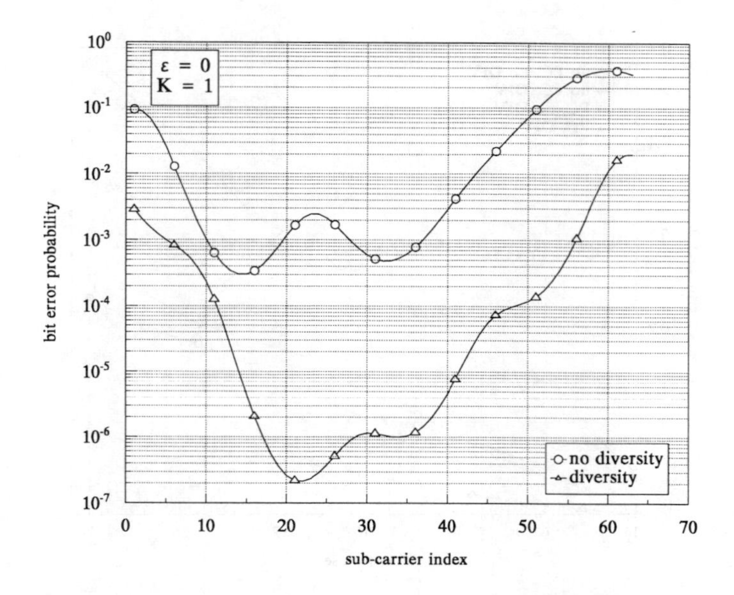

Figure 2: BER versus sub-carrier index with and without antenna diversity (one channel realization, SNR/bit = 10 dB, $K = 1$ active user, $\varepsilon = 0$).

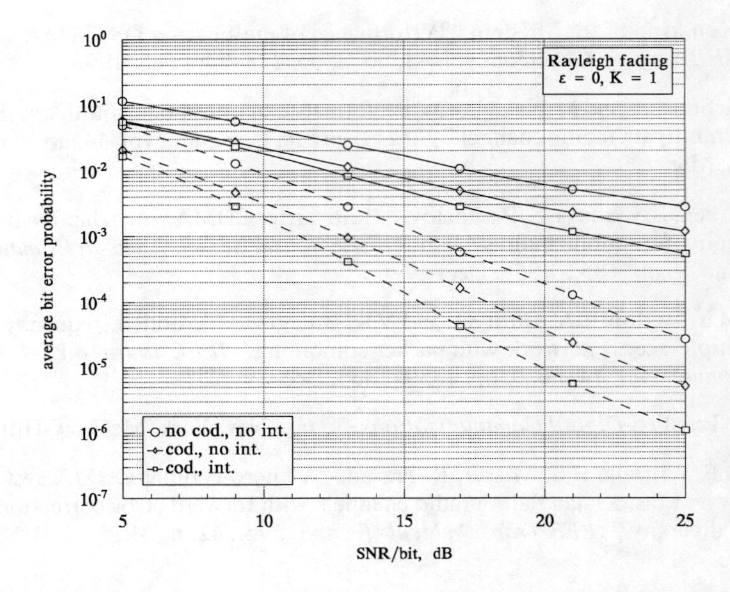

Figure 3: Average BER versus SNR/bit for different configurations of the receiver (solid line: no diversity, dashed line: diversity; K is the number of active users.).

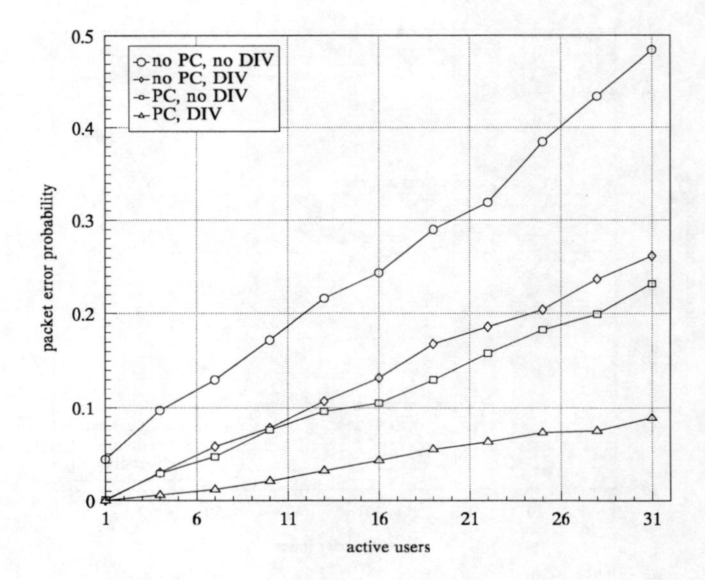

Figure 4: Packet error probability with different system configurations (PC: packet combining, DIV: antenna diversity).

Uplink spectral efficiency of multicarrier joint detection code division multiple access based cellular radio systems

Friedbert Berens, Peter Jung and Jörg Plechinger
University of Kaiserslautern,
P.O. Box 3049, D–67653 Kaiserslautern, Germany

Abstract

Recently, multicarrier joint detection code division multiple access (MC/JD–CDMA) [1] has been introduced as a viable candidate for future cellular radio systems. In this paper, the uplink system performance in a macrocellular environment shall be discussed in terms of the spectral efficiency. It is shown that the spectral efficiency depends on the chosen power control scheme. Results are presented for two different slow power control schemes and speech transmission at 8 kbit/s per mobile. Favourable performance is found for cellular networks with cluster order three to four.

1 Introduction

Presently, many research efforts are directed towards the development of third generation wireless systems, known as UMTS (Universal Mobile Telecommunications System) in Europe. It is expected that UMTS will support coexistence with existing radio systems, multioperator scenarios and hierarchical cell

K. Fazel and G.P. Fettweis (eds.), Multi-Carrier Spread-Spectrum, 211-218.

structures. These requirements call for flexible frequency engineering capabilities which can be provided by multicarrier (MC) techniques. Recently, MC techniques have been studied as an extension of the well known direct sequence (DS) based joint detection code division multiple access (JD–CDMA) [1]. This extension led to the new radio interface concept termed MC/JD–CDMA. In this communication, the system performance of MC/JD–CDMA in a macrocellular environment is presented in terms of the uplink spectral efficiency. The chosen uplink parameters are those of [2] and will not be explained further.

The system performance, and therefore the uplink spectral efficiency η, depends on the chosen service and the associated quality of service (QoS) criterion. In this communication, speech transmission at 8 kbit/s per mobile is taken into account. The QoS criterion comprises a maximum delay of 40 ms, a maximum acceptable outage probability $P_{\mathrm{out}}^{\mathrm{G}}$ of 10^{-2} and a maximum acceptable bit error ratio $P_{\mathrm{b}}^{\mathrm{G}}$ of 10^{-3}. To meet the QoS criterion, convolutional coding with code rate 1/2 and an interleaving depth of four TDMA frames is applied.

2 Simulation method

To evaluate the system performance in a cellular network, the interference limited case with numerous mobiles and base stations must be taken into account. Since mobile radio systems cannot be fully characterized by purely analytical approaches, the performance evaluation is based on Monte–Carlo simulation.

The efficient simulation exploits the fact that the time scale of the slow lognormal fading caused by shadowing is about three orders of magnitude larger than that of the fast Rayleigh fading associated with multipath propagation. Rayleigh fading becomes obvious when the mobiles cover a few wavelengths, i.e. several ten centimeters in the case of UMTS. Generally, the shadowing does not change over such short distances. The shadowing situation changes, however, when the mobiles move at least several ten meters. In the latter case, lognormal fading becomes apparent. In order to evaluate the effect of both kinds of fading, the simulation is carried out in two simulation steps.

In the first simulation step, only motions over a few wavelengths are considered. Hence, the Rayleigh fading is explicitly taken into account in the evaluation of the communication links whereas the shadowing is assumed to be constant. Owing to the displacement of the mobile locations, each link between a mobile and a base station has a particular received signal level attenuation caused by the shadowing. In an isolated cell of a cellular network where all mobiles are allocated to a single base station, a slow power control is feasible which equalizes the differences between the aforementioned received signal level attenuations. In the first simulation step which is called link level simulation, an isolated cell with perfect slow power control is therefore considered. The effect of the fast fading is however not combatted by any power control [1]. The isolated cell is loaded with $1 \leq \bar{K} \leq 8$ mobiles per frequency channel and time slot, cf. e.g. [1]. In the link level simulation, the C/I which is required to meet $P_{\mathrm{b}}^{\mathrm{G}} = 10^{-3}$ is determined. In this communication, antenna diversity is not considered. In the link level simulation, the intracell interference is taken into account.

The effect of intercell interference on the communication links in the isolated cell is dominated by shadowing. Therefore, the Rayleigh fading of the intercell interference can be neglected. Since motions over a few ten centimeters do not lead to noticeable changes of the shadowing, the intercell interference is modelled as stationary white noise in the link level simulation.

In the second simulation step called system level simulation, the effect of the intercell interference on the communication links of the aforementioned isolated cell is quantified by considering the shadowing explicitly. In the system level simulation, the fast Rayleigh fading, which was already taken into account in the link level simulation, is neglected. The system level simulation aims at the determining of the maximally allowed C/I to meet $P_{\mathrm{out}}^{\mathrm{G}} = 10^{-2}$ by simulating a cellular network which consists of several tens of cells, each being loaded with the same number $1 \leq \bar{K} \leq 8$ of mobiles per frequency channel and time slot. The allocation of a mobile to a particular cell is assumed to be seemless and infinitely fast. Therefore, the simulation yields upper performance bounds. Each link between a mobile and a base station is associated with a path loss exponent $\alpha = 4$ and slow fading with standard deviation $\sigma_{\mathrm{a}} = 8$ dB. The level of intercell interference also depends on the number of cells per cluster termed cluster order r. The aforementioned slow power control has been realized by two different algorithms. The first one controls the transmit powers in a cell in such a way that the received powers at the base station of that cell are always equal ("C = const."). The second one realizes the network wide centralized power control [3] striving for a C/I balancing ("C/I = const.").

The aforementioned isolated cell of the link level simulation is termed reference cell. Owing to the joint detection the reference cell does not have any intra-cell interference in the network simulation. To suppress boundary effects, the reference cell is located in the center of the simulated network area.

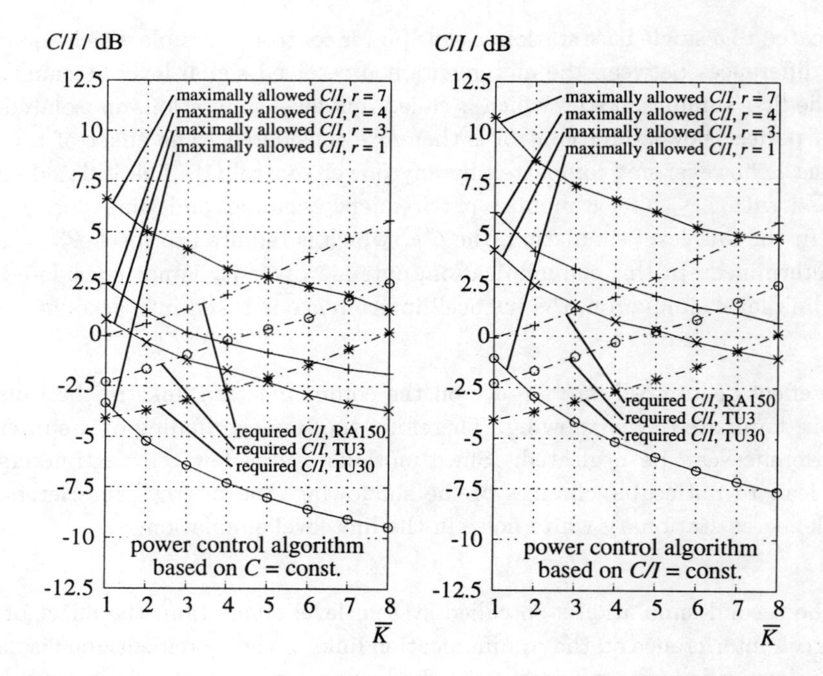

Figure 1. Carrier–to–interference ratio C/I in dB versus the mean number of mobiles $\bar{K}$ per frequency channel, time slot and cell

3 Maximally allowed C/I and required C/I

In Fig. 1 the carrier–to–interference ratio C/I in dB is depicted versus the mean number of mobiles $\bar{K}$ per frequency channel, time slot and cell. To accomodate as many mobiles as possible, the simultaneous maximization of $\bar{K}$ and minimization of r are strived for. Increasing $\bar{K}$ causes increasing interference which yields graceful degradation. This leads to a monotonically increasing C/I, required to meet $P_{\mathrm{b}}^{\mathrm{G}} = 10^{-3}$.

In Fig. 1, COST 207 typical urban area with mobile speed 3 km/h (TU3), COST 207 typical urban area with mobile speed 30 km/h (TU30), and COST 207 rural area with mobile speed 150 km/h (RA150) are considered as environment types. Due to improved time diversity, the performance for TU30 is better than for TU3. RA150 has the worst performance because of its low diversity potential.

For a given r, increasing $\bar{K}$ leads to an increasing intercell interference. For given r and C/I, the growing intercell interference yields an increase of the outage probability. In order to meet $P_{\text{out}}^{\text{G}} = 10^{-2}$, the maximally allowed C/I must therefore be decreased when $\bar{K}$ is increased [4]. Hence, the curves associated with the maximally allowed C/I are monotonically decreasing with increasing $\bar{K}$. The steepness of the decrease reduces for growing $\bar{K}$ which is a consequence of growing interferer diversity. The effect of the intercell interference diminishes with increasing r. Therefore, the maximally allowed C/I increases with increasing r for a given $\bar{K}$. The more sophisticated centralized power control allows higher values of the maximally allowed C/I for given r and $\bar{K}$ than the less expensive power control based on C = const.

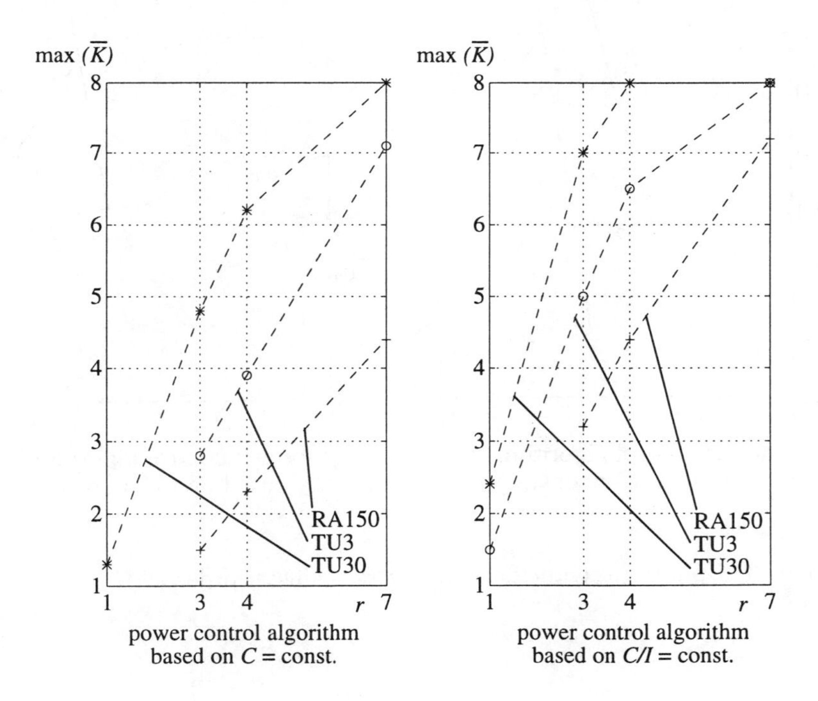

Figure 2. $\max(\bar{K}) \leq 8$ versus cluster order r

4 Uplink spectral efficiency

For a given environment type and a given $\bar{K}$, a cellular network with a particular cluster order r can be realized as long as the required C/I is lower than the

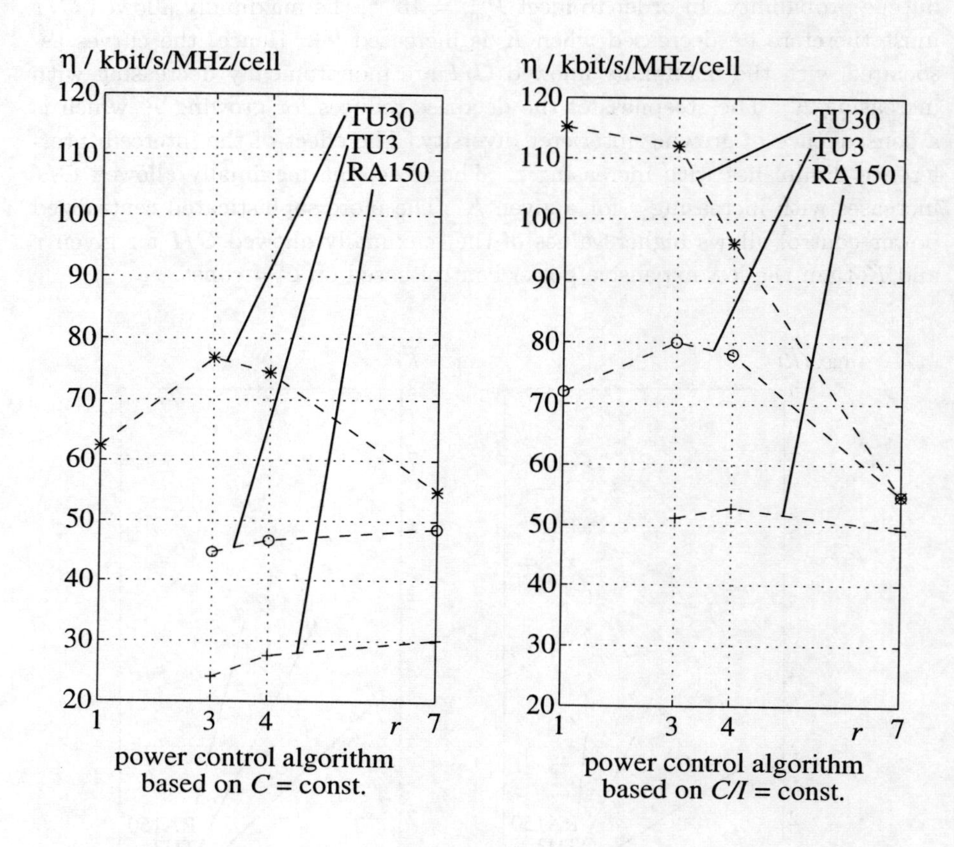

Figure 3. Uplink spectral efficiency η versus cluster order r

corresponding maximally allowed C/I. The minimum cluster order r_{min} and $\max(\bar{K}) \leq 8$ can thus be derived from Fig. 1 by determining the intersections of the curves associated with the maximally allowed C/I with those associated with the required C/I. Arbitrary values of r are possible for TU30. However, RA150 never allows the implementation of $r = 1$ whereas $r = 1$ is feasible for TU3 only when centralized power control is used.

Fig. 2 shows the obtained results for $\max(\bar{K}) \leq 8$ versus r with the environment types as parameters. In all cases $\max(\bar{K}) \leq 8$ increases with increasing r owing to the more favourable intercell interference situation. As expected, TU30 has the largest $\max(\bar{K})$ whereas RA150 has the lowest for a given r. Furthermore, centralized power control facilitates larger $\max(\bar{K})$ for given r and environment type than the simpler power control based on C =const.

With $R = 8$ kbit/s, the number of time slots $N_{\text{Z}} = 12$ per TDMA frame, the user bandwidth $B_{\text{u}} = 2$ MHz, the uplink spectral efficiency is

$$\eta = \frac{\max(\bar{K})}{r_{\text{min}}} \cdot \frac{RN_{\text{Z}}}{B_{\text{u}}} = \frac{\max(\bar{K})}{r_{\text{min}}} \cdot 48\,\text{kbit/s/MHz}. \tag{1}$$

Fig. 3 shows the uplink spectral efficiency η. As expected, the largest η around 110 kbit/s/MHz/cell is achieved with centralized power control. Furthermore, the application of antenna diversity which will be considered in a future publication leads to remarkable improvements of η. Obviously, $r = 3 \cdots 4$ facilitate a most favourable performance in all considered environment types.

Acknowledgements

The authors are indebted to the supercomputer staff at the Regionales Hochschulrechenzentrum Kaiserslautern (RHRK) for their support. Furthermore, the authors acknowledge the support of P.W. Baier, J. Blanz, M. Doetsch and T. Bing during the preparation of this work.

References

[1] F. Berens, P. Jung: Uplink performance of multicarrier joint detection code division multiple access. *IEE Electronics Letters*, vol. 33 (1997), no. 4, pp. 274–275.

[2] M. Naßhan, P. Jung, A. Steil, P.W. Baier: On the effects of quantization, nonlinear amplification and band–limitation in CDMA mobile radio systems using joint detection. *Proceedings of the 5th Annual International*

Conference on Wireless Communications (Wireless'93), Calgary (1993), pp. 173–186.

[3] S.A. Grandhi, R. Vijayan, D.J. Goodman, J. Zander: Centralized power control in cellular radio systems. *IEEE Transactions on Vehicular Technology*, vol. VT–42 (1993), pp. 466–468.

[4] A. Steil: Spektrale Effizienz digitaler zellularer CDMA–Mobilfunksysteme mit gemeinsamer Detektion. Fortschrittberichte VDI, series 10, no. 437, Düsseldorf: VDI–Verlag, 1996.

SIMULATION OF A DSSS/MCM SYSTEM IN A DOPPLER SPREAD CHANNEL

Thad B. Welch and Rodger E. Ziemer[1]

University of Colorado at Colorado Springs
Electrical and Computer Engineering Department
Colorado Springs, Colorado, USA

ABSTRACT - There has been a great deal of recent interest in the use of direct sequence spread spectrum (DSSS) in combination with multicarrier modulation (MCM). Although the proposed system implementations in the literature vary, all of these systems share the functionality of DSSS, MCM, and some form of a branch combiner at the receiver. This paper investigates the performance of such a system transmitting through a Doppler spread channel. In fast fading the proposed system performance exhibits an irreducible bit error probability (BEP) even for the single user case. Dramatic performance improvements can be realized with the use of a maximal ratio combiner (MRC). Finally, we investigate the effect of a more realistic branch combiner which utilizes the output from a channel gain averager for the receiver combiner weights.

[1] Work of R.E. Ziemer supported in part by the Office of Naval Research under contract N00014-920-J01761/P00004

K. Fazel and G.P. Fettweis (eds.), Multi-Carrier Spread-Spectrum, 219-226.

I. INTRODUCTION - Several authors, for example [1] - [4], propose and analyze DSSS/MCM systems for use in a mobile radio environment. In these systems, if the bandwidth of a wideband DSSS system exceeds the coherence bandwidth, $(\Delta f)_c$, of the channel, the signal is severely distorted [5]. This type of channel is said to be frequency-selective and a RAKE receiver [6] can be used to enhance system performance [7]. If a single baseband DSSS signal is modulated by multiple carriers that are separated to prevent spectral overlap of the adjacent sub-carrier DSSS bandpass signals, these sub-carriers will experience independent fading. With multiple replicas of the message signal being transmitted through independent fading channels, diversity gain can be achieved by the receiver. An MRC can be used to recombine these message signals to improve the systems error performance [8].

Kondo and Milstein, [4], propose and analyze such a system and show that their DSSS/MCM/MRC system performance with M sub-carriers is identical to a single carrier RAKE receiver system occupying the same bandwidth and processing M resolvable paths ($L_1 = M$). Their system controls the number of resolvable paths available to the DSSS/MCM/MRC receiver by varying the chip period and ensuring a set of provided inequalities are satisfied. This procedure provides only one resolvable path for each sub-carrier's channel. If each sub-carrier has bandwidth $BW_M \geq (\Delta f)_c$, each channel will fade independently and have no selectivity [4]. To maintain a constant system bandwidth, the product of the chip period and processor gain must remain constant. As we add more carriers, the chip period increases and the processor gain decreases. This system was granted a U.S. Patent [9] and will be the baseline design which we study to give improved reception through a fast fading channel.

While the longer chips associated with Kondo and Milstein's DSSS/MCM/MRC system help to mitigate the delay spread/fading problem, the system is more susceptible to the effects of Doppler spread. Doppler spread will cause amplitude and phase fluctuations of the received message chips. These amplitude and phase fluctuations result in an irreducible bit error probability (BEP) even for a single user system.

II. SYSTEM MODEL - Our system transmitter/channel block diagram is shown in Figure 1. Differential phase shift keying (DPSK) is used to help reduce the effect of the Doppler-induced rapid phase variations of the received signal. The DSSS spreading process is shown as the multiplication of the DPSK data with the local pseudo-noise (PN) code. The multicarrier local oscillators are at frequencies $f_1, f_2 \ldots f_M$, where M is the total number of sub-carriers. RW_k is the complex multiplicative Rayleigh weight that models the independent fading associated with the *k-th* sub-carrier channel (k = 1...M) and I&Q AWGN are the in-phase and quadrature components of an additive white Gaussian noise process.

Our receiver block diagram is shown in Figure 2. The DSSS despreading process is shown as the multiplication of the received signal with the local estimate of the PN code. We will assume perfect code tracking. I&D represents the receiver integrate and dump function which is synchronized to the bit transition times. The delay* blocks represent the delay and multiply associated with the DPSK decode process and * represents complex conjugation. $\left|\overline{RW_k}\right|$ represents the combiner weight associated with the *k-th* branch. This model can be expanded to accommodate additional sub-carriers by adding more branches.

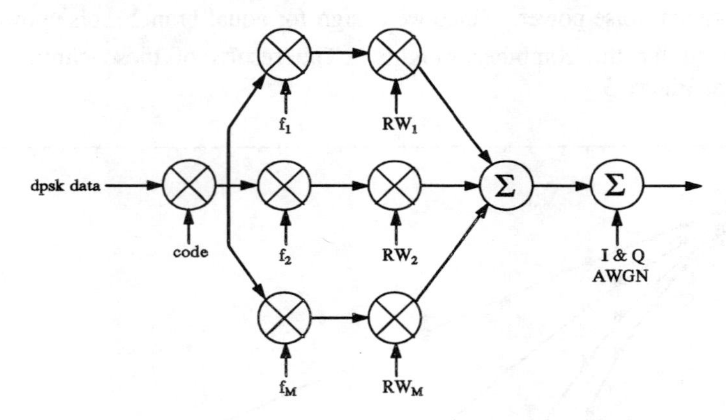

Figure 1. Transmitter and Channel Block Diagram

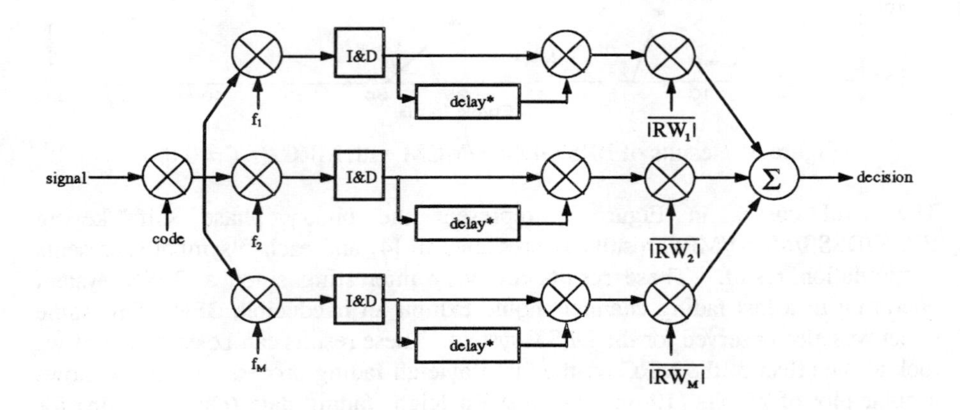

Figure 2. Receiver Block Diagram

III. SIMULATION RESULTS - We wrote a series of MATLAB® m-files that implemented the block diagrams shown in Figure 1 and 2. The complex multiplicative Rayleigh weights, RW_k, are generated by a channel simulator using an infinite impulse response (IIR) digital filter [10].

The parameter f_dT [11] is used to characterize the value of the Doppler spread experienced by the signal, where f_d is the maximum Doppler shift (in hertz) and T is the symbol period (in seconds). The first series of simulations implements the Figure 1 and 2 systems with the exception that a perfect MRC is used for the branch combiner. For this ideal case the branch combiner weights should be g_k^*/N_k [8], where g_k^* is the conjugate of the *k-th* branch channel gain and N_k is the *k-th* branch noise power. Since we design for equal branch noise powers, g_k^* will be used for the combiner weights. The results of these simulations are provided as Figure 3.

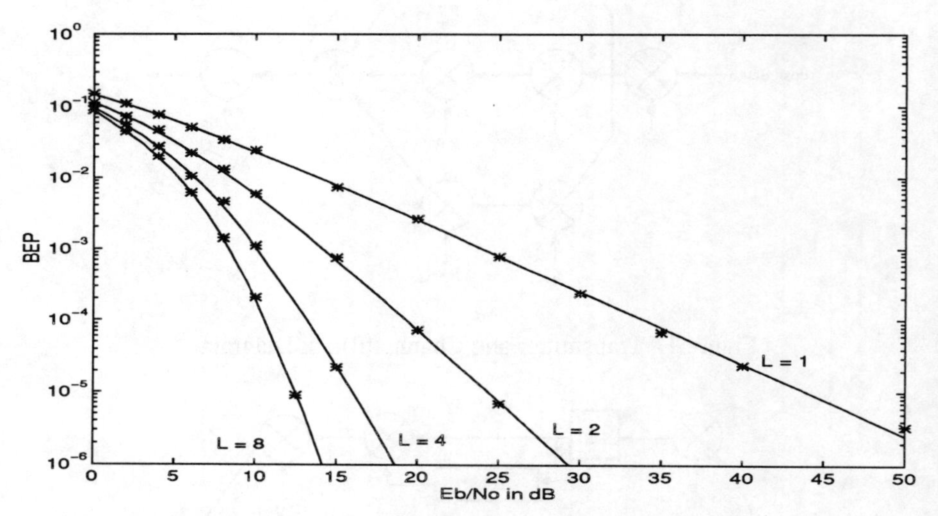

Figure 3. Results of BPSK/DSSS/MCM with MRC, (f_dT = 0.02).

The solid curves in Figure 3 represent the binary phase shift keying BPSK/DSSS/MCM/MRC results as presented in [4] and each * symbol represents a simulation result. These results are very interesting since a BPSK system operating in a fast fading channel should exhibit an irreducible BEP. This same effect was also observed for the DPSK system. These results can be explained if we look at the effect of the MRC on the fast Rayleigh fading process. Figure 4 shows a polar plot of 24 bits (10 msec) of fast Rayleigh fading data (channel gain) for f_dT = 0.08. This large value of f_dT is used only for illustrative purposes since it helps to separate the plotted samples of the channel gain. The *'s on the polar plot indicate the channel gains associated with the middle of a single bit.

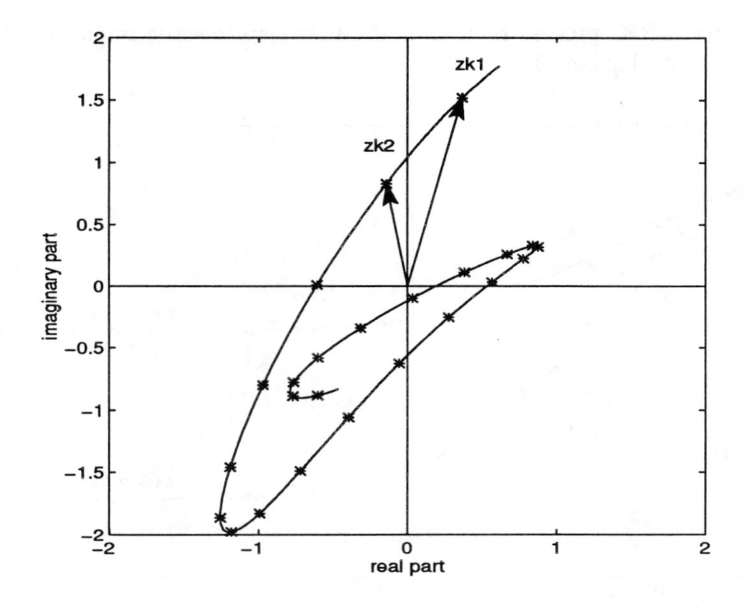

Figure 4. Typical Fast Rayleigh Fading Channel Gains for 24 Data Bits, $f_d T = 0.08$.

If we denote the received signal in the *k-th* branch associated with the present bit as z_{k1} and the previous bit as z_{k2} the decision statistic for this DPSK system is $\mathrm{Re}\{z_{k1} z_{k2}^{*}\}$ [8]. If the AWGN in the system is assumed to be zero, and we account for the phase shift associated with the modulation process, as $E_b/N_0 \to \infty$, then $z_{k1} = g_{k1}$ and $z_{k2} = g_{k2}$. If we isolate 2 channel gain bits, as shown in Figure 4, and labeled these bits z_{k1} and z_{k2}, the effect of multiplying these vectors by the complex conjugates of the respective channel gains, g_{k1}^{*} and g_{k2}^{*}, results in a mapping of these vectors to the real axis. This same mapping process also occurs in a BPSK system. Therefore, the effect of fast Rayleigh fading is "undone," given perfect MRC (perfect channel gain estimation for both the magnitude and phase components), and the system performs as if it were operating in a flat fading channel.

Since perfect MRC is not possible for every bit, a channel gain estimator, as shown in Figure 2, will be needed. Figure 5 shows the result of assuming the magnitude of the channel gain associated with the middle of a bit is known and used as the branch weight. This branch weight is approximately the average of the magnitude of the channel gain over the entire bit. This averaging procedure, and the extension of the procedure where the magnitude of the channel gains associated with an integer number of bits will be called average channel gain combining

(ACGC). This ACGC process represents the first step towards implementing a stochastic gradient adaptive filter.

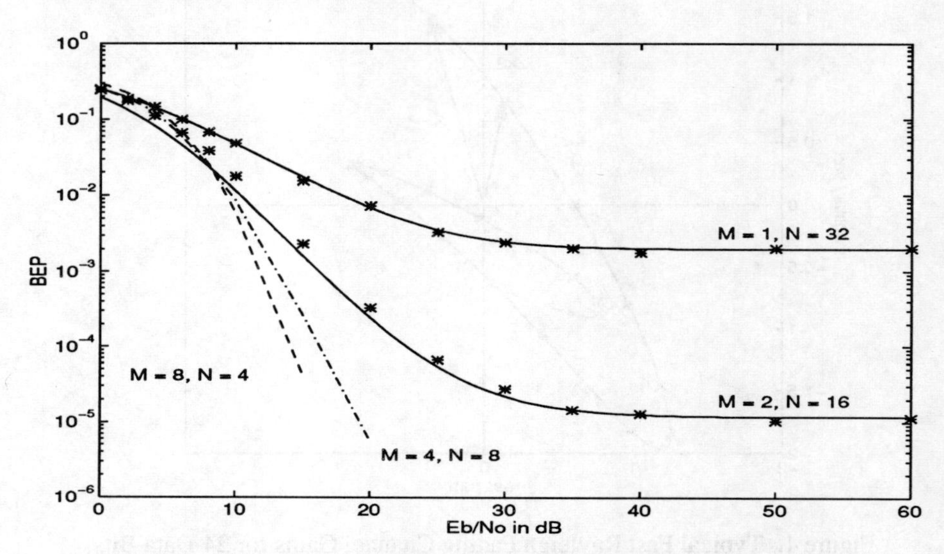

Figure 5. Results for DPSK/DSSS/MCM with ACGC, $f_dT = 0.02$.

The 2 solid curves in Figure 5 and 6 represent the closed form expression for a DPSK system operating in a Doppler spread channel without interference [12] and [13]. As expected, as f_dT increases the system performance degrades, Figure 6.

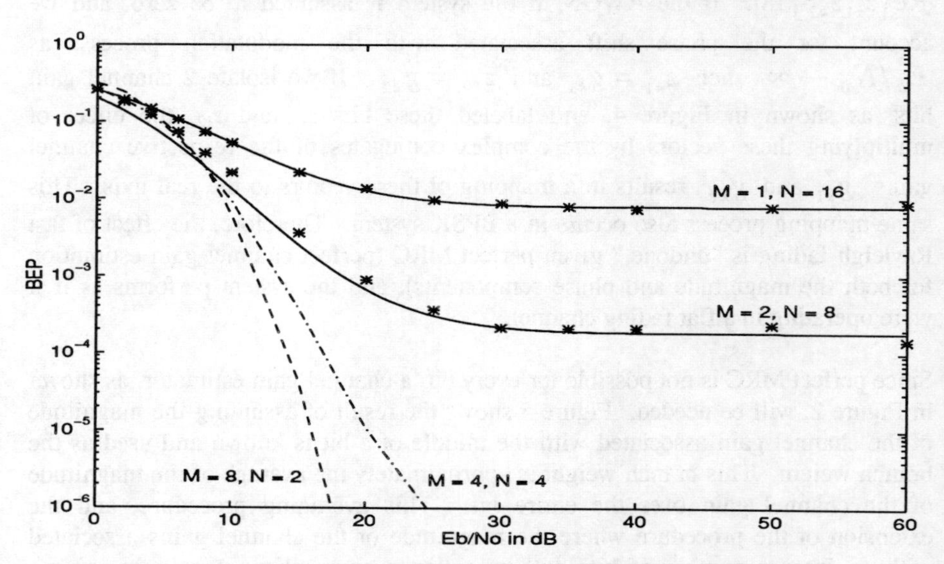

Figure 6. Results for DPSK/DSSS/MCM with ACGC, $f_dT = 0.04$.

The effect of varying the ACGC averaging interval is investigated next. For this simulation an averaging interval of 1, 4, 16 and 64 bits is used. For comparison, the equal gain combiner (EGC) case is also calculated. The results of these simulations for $M = 2$ and $M = 4$ are shown in Figure 7.

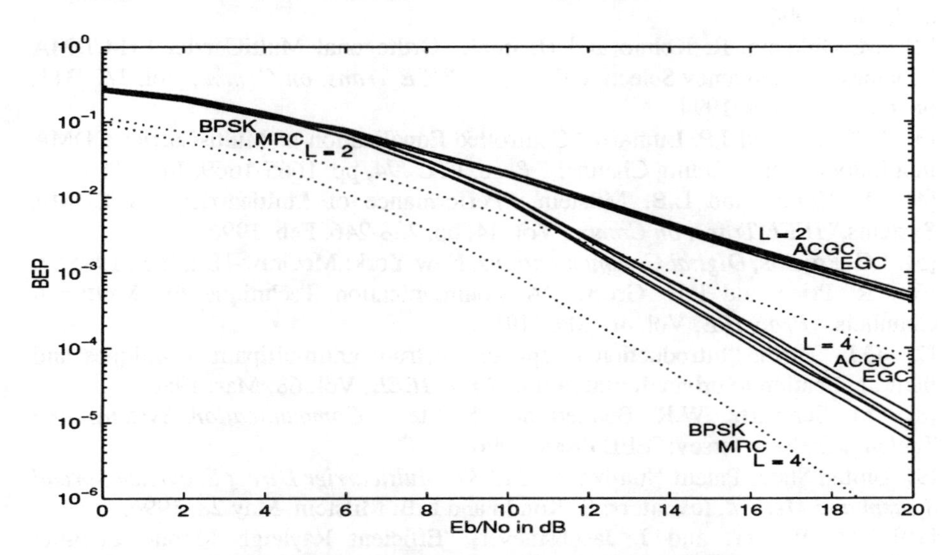

Figure 7. Results for DPSK/DSSS/MCM with ACGC and EGC ($M = 2$ and 4).

The general trend is towards degraded performance as the averaging interval increases from 1 to 64 bits. The performance of the EGC system is better than the 64 bit averaging interval case but worse than the single bit averaging interval case. The dashed curves represent the performance of the BPSK/DSSS/MCM with MRC as shown in Figure 3.

IV. CONCLUSIONS - The DPSK/DSSS/MCM with branch combining is much more robust in the presence of fast fading than a single carrier ($M = 1$) system. The effects of the irreducible BEP associated with this channel can be reduced as more sub-carriers are added to the system. The MRC is much more effective at improving system performance than the ACGC. At low E_b/N_o values, the performance difference between the MRC and ACGC system is small but increases greatly as E_b/N_o increases. This is because the ACGC system will eventually reach an irreducible BEP; however, the MRC system will exhibit no such effect. If system specifications require operating below the irreducible bit error probability, a MRC is a very attractive *theoretical* solution to this problem. The pursuit of an improved branch combiner must be continued.

REFERENCES

[1] E. Sourour and M. Nakagawa, "Performance of Orthogonal Multi-Carrier CDMA in a Multipath Fading Channel," *IEEE Trans. on Comm.*, Vol. 44, pp. 356-367, Mar. 1996.

[2] M. Ohkama, R. Kohno and H. Imai, "Orthogonal MultiCarrier FH-CDMA Schemes for Frequency Selective Fading," *IEICE Trans. on Comm.*, Vol. J77-B11, pp. 691-702, Nov. 1994.

[3] R.N. Yee and J.P. Linnartz "Controlled Equalization of Multi-Carrier CDMA in a Indoor Rician Fading Channel," *Proc. VTC '94*, pp. 1665-1669, Jun. 1994.

[4] S. Kondo and L.B. Milstein, "Performance of Multicarrier DS CDMA Systems," *IEEE Trans. on Comm.*, Vol. 44, pp. 238-246, Feb. 1996.

[5] J.G. Proakis, *Digital Communications*, New York: McGraw-Hill, Inc., 1995.

[6] R. Price and P.E. Green, "A Communication Technique for Multipath Channels," *Proc. IRE*, Vol. 46, Mar. 1958.

[7] G.L. Turin, "Introduction to spread-spectrum antimultipath techniques and their application to urban digital radio," *Proc. IEEE*, Vol. 68, Mar. 1980.

[8] M. Schwartz, W.R. Bennett and S. Stein, *Communication Systems and Techniques*, New Jersey: IEEE Press, 1996.

[9] United States Patent Number: 5,521,937, *Multicarrier Direct Sequence Spread System and Method*, Inventors: S. Kondo and L.B. Milstein, May 28, 1996.

[10] M. Wickert and J. Jacobsmeyer, "Efficient Rayleigh Mobile Channel Simulation Using IIR Digital Filters," *Proc. ICSPAT '95*, pp. 391-395, Oct. 1995.

[11] K. Feher, Ed., *Advance Digital Communications: Systems and Signal Processing Techniques*, Colorado: Crestone Engineering, 1996.

[12] K. Hirrade, M. Ishizuka, F. Adachi and K. Ohtani, "Error-Rate Performance of Digital FM with Differential Detection in Land Mobile Radio Channels", *IEEE Trans. on Veh. Tech.*, Vol. 28, pp. 204-212, Aug. 1979.

[13] H. Voelcker, "Phase-Shift-Keying in Fading Channels", *Proc. IEE*, Vol. 107, part B, pp. 31-38, Jan. 1960.

ON THE PERFORMANCE OF ASYNCHRONOUS MULTI-CARRIER CDMA

H. OCHIAI and H. IMAI
Institute of Industrial Science, University of Tokyo,
7-22-1 Roppongi, Minato-ku, Tokyo 106, Japan (Tel: +81-3-3402-6231(Ex. 2327), Fax: +81-3-3402-7365, E-mail: ochiai@imailab.iis.u-tokyo.ac.jp)

Abstract. Performance of asynchronous MC-CDMA(OFDM-CDMA) over AWGN channel for uplink communications systems is studied. For a given bandwidth, it is shown that asynchronous MC-CDMA eventually outperforms asynchronous DS-CDMA with random sequences. We also discuss the synthesis of power efficient complementary sequences as the orthogonal sequences of an uplink MC-CDMA system.

1. Introduction

Multicarrier modulation or orthogonal frequency division multiplexing(OFDM), is a promising technique for mobile communications systems, since it has a strong immunity to multipath fading without employing complicated adaptive equalization. Recently, the novel combined scheme of OFDM and spread spectrum multiple access(SSMA), which is called multi-carrier code division multiple access(MC-CDMA) or OFDM-CDMA, has been proposed for indoor and mobile communications systems[1-3]. While a lot of contributions have been done for synchronous MC-CDMA, little attention has been paid for the asynchronous case, because in the uplink, orthogonality of the spreading codes may be severely distorted, causing a strong multiple-access interference(MAI).

On additive white Gaussian noise(AWGN) channel, uplink systems largely have two major requirements which are of critical importance; synchronization and power efficiency.

In this paper, performance of MC-CDMA with coherent detection for an uplink CDMA system over AWGN channel is discussed, and computer simulations are presented.

2. Uplink MC-CDMA System Model

Let us consider multiple access system with K active users. For simplicity, each user is assumed to transmit one data symbol per each OFDM symbol.

K. Fazel and G.P. Fettweis (eds.), Multi-Carrier Spread-Spectrum, 227-234.

MC-CDMA transmitter of the k-th user is shown in Fig. 1(a), where $b^{(k)}$ and $C_l^{(k)}$ denote, respectively, the information symbol and the l-th spreading code of length N of the k-th user, where $k = 0, 1, \ldots, K-1$ and $C_l^{(k)} \in \{-1, +1\}$, for $l = 0, 1, \ldots, N-1$ and $b^{(k)} \in \{-1, +1\}$, if BPSK is applied. By using N orthogonal codes, the maximum number of users K is equal to N.

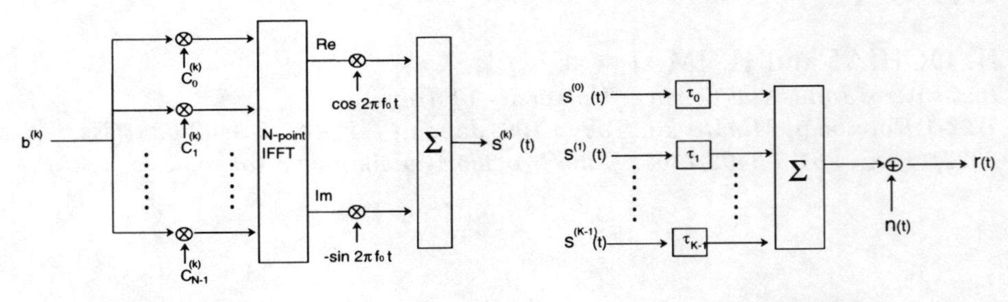

(a) Transmitter for the k-th user (b) Channel model

Fig.1. Uplink MC-CDMA system model

The complex baseband signal for the m-th symbol of the k-th user can be written as

$$s_m^{(k)}(t) = \sqrt{\frac{E_s}{NT_s}} \cdot b_m^{(k)} \sum_{l=0}^{N-1} C_l^{(k)} e^{j2\pi \frac{l}{T_s} t} f(t - mT_s), \tag{1}$$

where $f(t)$ is the pulse waveform of each symbol, given as

$$f(t) = \begin{cases} 1, & 0 \leq t < T_s, \\ 0, & \text{otherwise,} \end{cases}$$

T_s is a symbol period and E_s is the transmitted energy per symbol. A complex baseband signal notation will be used throughout the paper. To simplify our argument, neither the guard interval nor the influence of intersymbol interference(ISI) of each user will be taken into account.

With a perfect power control, the received signal at the base station is given by(see Fig.1(b))

$$\begin{aligned} r(t) &= \sum_{m=-\infty}^{\infty} \sum_{k=0}^{K-1} s_m^{(k)}(t - \tau_k) + n(t) \\ &= \sum_{m=-\infty}^{\infty} s_m^{(0)}(t) + \sum_{m=-\infty}^{\infty} \sum_{k=1}^{K-1} s_m^{(k)}(t - \tau_k) + n(t) \end{aligned} \tag{2}$$

where $n(t)$ denotes the additive Gaussian noise with one-sided power spectral density N_0 and τ_k is the delay of the k-th user. Without loss of generality, we have assumed that $\tau_0 \equiv 0$ and $0 \leq \tau_k < T_s$ for $k = 1, 2, \ldots, K-1$, and τ_k can be considered as independent, identically distributed(iid) random variables.

The output of the correlator matched to the m-th symbol of the 0-th user can be written as

$$z_m^{(0)} = \sqrt{E_s} b_m^{(0)} + \sqrt{\frac{E_s}{N}} \sum_{k=1}^{K-1} I_{m-1}^{(k)} + \sqrt{\frac{E_s}{N}} \sum_{k=1}^{K-1} I_m^{(k)} + \mathcal{N}, \tag{3}$$

where $\mathcal{N}$ denotes Gaussian noise term, and $I_{m-1}^{(k)}$ and $I_m^{(k)}$ are MAI terms which are, respectively, given by

$$I_{m-1}^{(k)} = b_{m-1}^{(k)} \sum_{n=0}^{N-1} \sum_{l=0}^{N-1} C_l^{(0)} C_n^{(k)} e^{-j2\pi \frac{l}{T_s} \tau_k} \frac{1}{T_s} \int_0^{\tau_k} e^{j2\pi \frac{(l-n)}{T_s} t} dt \tag{4}$$

and

$$I_m^{(k)} = b_m^{(k)} \sum_{n=0}^{N-1} \sum_{l=0}^{N-1} C_l^{(0)} C_n^{(k)} e^{-j2\pi \frac{l}{T_s} \tau_k} \frac{1}{T_s} \int_{\tau_k}^{T_s} e^{j2\pi \frac{(l-n)}{T_s} t} dt. \tag{5}$$

2.1. Performance of Synchronous MC-CDMA

Performance evaluation and interference calculations of synchronous MC-CDMA applied to uplink over Rayleigh fading channel have been done in [4], where a perfect synchronization was assumed. With perfect synchronization, i.e., $\tau_k = 0$ for all k, Eqs.(4-5) can be respectively rewritten as

$$I_{m-1}^{(k)} = 0 \qquad \text{and} \qquad I_m^{(k)} = \sum_{l=0}^{N-1} C_l^{(0)} C_l^{(k)} = 0,$$

since we use the orthogonal codes for each user. Consequently, MAI-free communication can be achieved. However, as briefly discussed in [3], for practical uplink systems, perfect synchronization is not feasible, and this leads to the significant performance degradation due to the MAI. One of the methods to alleviate the synchronization restriction is to choose T_s as large as possible, or equivalently, increase the number of carriers for a given bandwidth. The influence of synchronization error will be studied later by computer simulations.

2.2. Performance of Asynchronous MC-CDMA

Performance of asynchronous MC-CDMA has been evaluated recently in [5], based on the method of Pursley[6]. However, the result given in [5] may be underestimated(too optimistic). In what follows, we will take different and simpler approach to derive the bit error performance.

Let us consider Eq.(2) again, where $\tau_0 \equiv 0$, and for $k = 1, 2, \ldots, K-1$, τ_k are iid random variables uniformly distributed over $[0, T_s]$. Let us also assume that $\mathcal{S}_k \equiv s_m^{(k)}(t-\tau_k)$, where $k = 1, 2, \ldots, K-1$, are iid random variables. Since multicarrier signal consists of sum of N waveforms with variance $\sigma^2 = (\sqrt{E_s/NT_s})^2$, as N increases, its waveform will have a Gaussian distribution with zero mean and variance $N\sigma^2$. That is, for asynchronous MC-CDMA system, interference from other users can be well approximated as a Gaussian noise with zero mean and variance E_s/T_s. Hence, the total density of noise, i.e., $\sum_{k=1}^{K-1} \mathcal{S}_k + \mathcal{N}$, is found to be

$$N' = (K-1) \cdot \frac{E_s}{T_s \cdot W} + N_0 = (K-1) \cdot \frac{E_s}{N+1} + N_0, \tag{6}$$

where $W = (N+1)/T_s$ is the total (one-sided) bandwidth.

Bit error probability can be found as follows:

$$\text{BPSK: } P_e(MC) = Q\left(\sqrt{\frac{E_s}{N'/2}}\right) = Q\left(\left\{\frac{K-1}{2(G_{MC}+1)} + \frac{N_0}{2E_b}\right\}^{-\frac{1}{2}}\right) \tag{7}$$

$$\text{QPSK: } P_e(MC) = Q\left(\sqrt{\frac{E_s}{N'}}\right) = Q\left(\left\{\frac{K-1}{G_{MC}+1} + \frac{N_0}{2E_b}\right\}^{-\frac{1}{2}}\right), \tag{8}$$

where E_b denotes the energy per transmitted bit, and $G_{MC} \equiv N$ in the equations above denotes the number of carriers, which corresponds to the processing gain. $Q(x)$ is given by $Q(x) = \frac{1}{\sqrt{2\pi}} \int_x^\infty \exp\left(-\frac{u^2}{2}\right) du$.

Based on the standard Gaussian approximation, on the other hand, bit error rate of DS-CDMA with random sequences of rectangular pulse is given as [6]:

$$\text{BPSK: } P_e(DS) \approx Q\left(\left\{\frac{K-1}{3G_{DS}} + \frac{N_0}{2E_b}\right\}^{-\frac{1}{2}}\right) \tag{9}$$

$$\text{QPSK: } P_e(DS) \approx Q\left(\left\{\frac{2(K-1)}{3G_{DS}} + \frac{N_0}{2E_b}\right\}^{-\frac{1}{2}}\right) \tag{10}$$

Since the spectra of MC-CDMA signal are allocated to overlap, for the same given bandwidth and symbol period, processing gain of MC-CDMA is twice as much as DS-CDMA[7], i.e., $G_{MC} \approx 2 \cdot G_{DS}$. Substituting this to Eq.(7-10), for a given bandwidth W and K, we can obtain

$$P_e(MC) < P_e(DS),$$

which indicates that when comparing under the same bandwidth and without co-channel interference cancellation, MC-CDMA eventually outperforms DS-CDMA with random sequences.

3. Orthogonal Set of Power Efficient Sequences for Uplink Channel

Although OFDM is spectrally efficient, it is known to be power inefficient, because its waveform tends to have a large peak to average power ratio(PAPR). For cellular systems, especially uplink, in order to use the power of transmitter efficiently, class C or other nonlinear power amplifiers are typically enployed. These kinds of amplifiers, however, cause a nonlinear distortion which severely degrades the performance of multicarrier systems.

By using complementary pair of sequences[10] for spreading codes of MC-CDMA, PAPR of multicarrier waveform can be maintained below 3dB[8]. In [9], complementary pair of sequences were proposed for MC-CDMA, and their property after pulse shaping was intensively examined. However, no comment on their orthogonality was given. In MC-CDMA systems, Walsh-Hadamard(W-H) sequences are typically chosen for orthogonal sequences. These sequences, however, tend to yield a high PAPR, because of their particular autocorrelation property. An orthogonal set of complementary sequences can readily be found, therefore it may be preferable to utilize the orthogonal set of complementary sequences for uplink MC-CDMA. Note that for the downlink, however, since sum of several orthogonal codes are transmitted, complementary property will not be preserved as long as $K \geq 2$. In that case, W-H sequences are suitable[11].

3.1. Synthesis of Orthogonal Set of Complementary Sequences

There are several methods to construct an orthogonal set based on complementary sequences, and in what follows, we give a simple example. Let $N = 2^m$, $m = 1, 2, \ldots$, denote the number of carriers and $\boldsymbol{A}_N$ and $\boldsymbol{B}_N$ be $N \times N/2$ matrices which can be recursively given as

$$\boldsymbol{A}_N = \begin{bmatrix} \boldsymbol{A}_{N/2} & \boldsymbol{B}_{N/2} \\ \boldsymbol{A}_{N/2} & \boldsymbol{B}_{N/2} \end{bmatrix}, \text{ and } \boldsymbol{B}_N = \begin{bmatrix} \boldsymbol{A}_{N/2} & \overline{\boldsymbol{B}}_{N/2} \\ \overline{\boldsymbol{A}}_{N/2} & \boldsymbol{B}_{N/2} \end{bmatrix},$$

where $\boldsymbol{A}_2 = \begin{bmatrix} + \\ + \end{bmatrix}$, $\boldsymbol{B}_2 = \begin{bmatrix} + \\ - \end{bmatrix}$, and $\overline{\boldsymbol{X}}$ indicates the reversing of all the elements of $\boldsymbol{X}$. Then, we can construct the $N \times N$ matrix, each row of which is composed of complementary pair of sequences orthogonal to other rows by the concatenation of $\boldsymbol{A}_N$ and $\boldsymbol{B}_N$, namely, $\boldsymbol{H}_N = \begin{bmatrix} \boldsymbol{A}_N & \boldsymbol{B}_N \end{bmatrix}$.

Example: We can obtain an orthogonal set of complementary sequences of length four as follows.

$$\boldsymbol{H}_4 = \begin{bmatrix} \boldsymbol{A}_4 & \boldsymbol{B}_4 \end{bmatrix} = \begin{bmatrix} \boldsymbol{A}_2 & \boldsymbol{B}_2 & \boldsymbol{A}_2 & \overline{\boldsymbol{B}}_2 \\ \boldsymbol{A}_2 & \boldsymbol{B}_2 & \overline{\boldsymbol{A}}_2 & \boldsymbol{B}_2 \end{bmatrix} = \begin{bmatrix} + & + & + & - \\ + & - & + & + \\ + & + & - & + \\ + & - & - & - \end{bmatrix}$$

We can readily prove that each row of the matrix above satisfies the condition for Golay's complementary sequence[10].

4. Simulation Results

In this section, performance of the system considered above is evaluated by computer simulations. PAPR's of baseband multicarrier signals based on complementary and W-H orthogonal sequences are shown in Fig.2. The number of carriers is chosen to be 64. In general, W-H sequence yields PAPR as much as $10\log_{10} N = 18$ dB. Even if we avoid using these sequences which yield maximum PAPR, W-H sequences tend to yield relatively high PAPR as shown in Fig.2. On the other hand, by using complementary sequences, PAPR can be successfully reduced to less than or equal to 3 dB.

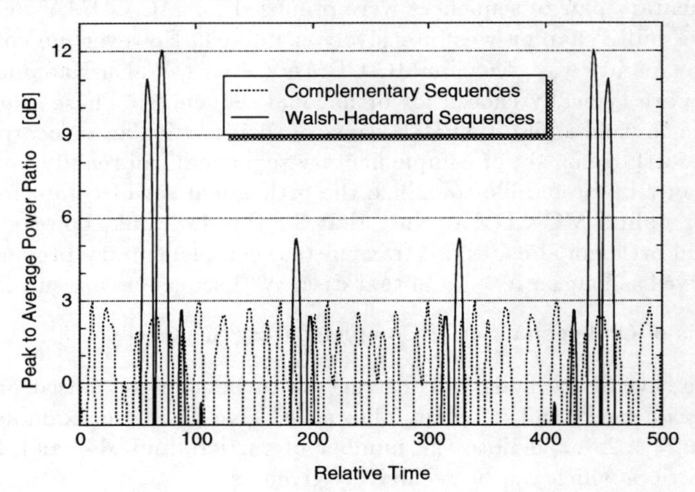

Fig.2. PAPR of MC-CDMA with orthogonal sequences of length $N = 64$

Shown in Fig.3 is the bit error performance of asynchronous MC-CDMA with $N = 64$. For the spreading codes in the simulation, we have used power efficient complementary sequences. Theoretical values obtained from Eqs.(7-8), agree with the simulation results especially for large number of users K. For small K, however, theoretical values may seem to be too optimistic. This may be because the waveforms based on complementary sequences are PAPR restricted, and therefore cannot be Gaussian. For large K, on the other hand, central limit theorem holds and they approach a Gaussian.

As discussed in Section2, synchronous MC-CDMA will be very sensitive to synchronization errors due to the destruction of orthogonality. Let us assume that some synchronization error exists, i.e., τ_k are uniformly distributed over $[0, \beta T_s]$ where β is the synchronization error normalized by the symbol period and therefore $0 < \beta \leq 1$. $\beta = 0$ corresponds to the perfect synchronization, while $\beta = 1$ is considered to be the asynchronous system which we have just examined.

Shown in Fig.4 is the bit error performance of a 32-carrier MC-CDMA system with BPSK, where $K = 4, 8, 16$ and 32, and $E_b/N_0 = 6$ dB. Theoretical values of both synchronous and asynchronous cases are also given for comparison. From the figure, it can be seen that in order to avoid the considerable degradation due to MAI, β must be less than 0.01. It is also interesting to note that for synchronous systems with relatively small K, once β exceeds some threshold, the overall performance will be inferior to asynchronous systems. This may be because MAI cannot be assumed as Gaussin noise, and therefore severely interferes the users.

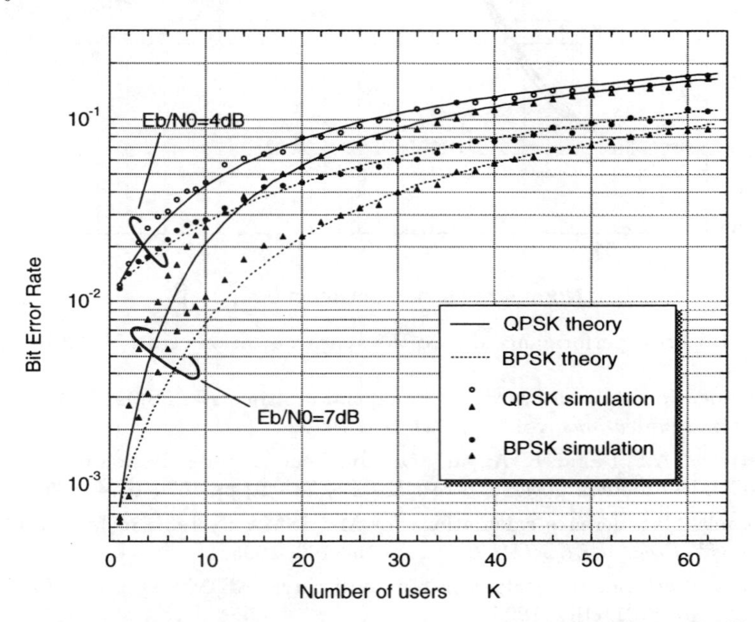

Fig.3. Bit error performance of asynchronous MC-CDMA with $N = 64$

5. Conclusions

The performance of asynchronous(uplink) MC-CDMA was studied. By considering its frequency efficiency, it has been shown that MC-CDMA is comparable to DS-CDMA. Moreover, with precise synchronization and orthogonal codes, performance can be significantly improved. Complementary sequences can be also used to mitigate the PAPR of multicarrier signals.

References

1. N. Yee, J.P. Linnartz, G. Fettweis, "Multi-Carrier CDMA in Indoor Wireless Radio Networks," *IEICE Trans. Commun.*, vol.E77-B, no.7, pp.900-904, July 1994.
2. K. Fazel, S.Kaiser, M.Schnell, "A flexible and high performance cellular mobile

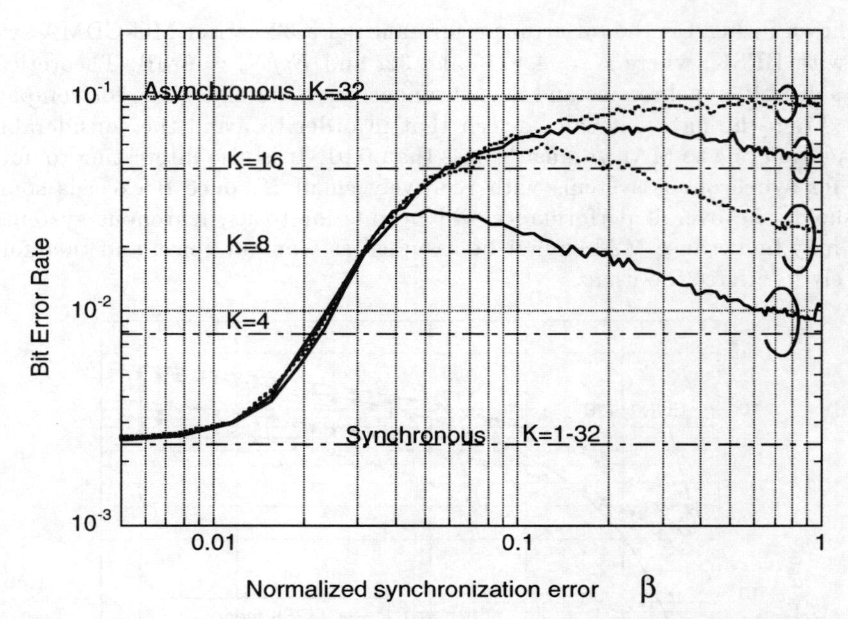

Fig.4. Bit error performance due to the synchronization error(N=32,BPSK)

communications system based on orthogonal multi-carrier SSMA," *Wireless Personal Communications*, vol.2, pp.121-144, 1995.

3. G. Fettweis, A.S. Bahai, K. Anvari "On multi-carrier code division multiple access (MC-CDMA) modem design," *Proc. IEEE VTC '94*,pp.1670-1674, 1994.
4. M.Schnell, "Interference calculations for MC-SSMA Systems in Mobile Communications," *Proc. IEEE PIMRC '95*, pp.158-163, 1995.
5. T.F. Ho, "Performance evaluation for multi-carrier CDMA system," *Proc. IEEE VTC '96*, pp.1101-1105, 1996.
6. M.B. Pursley, "Performance evaluation for phase-coded spread-spectrum multiple-access communication–part I: system analysis," *IEEE Trans.Commun.*, vol.COM-25, No.8, pp.795-799, August 1977.
7. N. Morinaga, M. Nakagawa, R. Kohno, "New concepts and technologies for achieving highly reliable and high-capacity multimedia wireless communications systems," *IEEE Commun. Mag.*, pp.34-40, January 1997.
8. B.M. Popović, "Synthesis of power efficient multitone signal with flat amplitude spectrum," *IEEE Trans.Commun.*, vol.COM-39, No.7, pp.1031-1033, July 1991.
9. T.F. Ho, V.K. Wei, "Synthesis of low-crest waveforms for multicarrier CDMA system," *Proc. IEEE GLOBECOM '95 Mini-Conference*, pp.131-135, 1995.
10. M.J.E. Golay, "Complementary series," *IRE Trans. Inform. Theory*, vol.IT-7, pp.82-87, April 1961.
11. L. Freiberg, V.K. Bhargava, "Efficient use of orthogonal codes in multi-carrier code division multiple access mobile cellular communications systems," *Proc. WMWCC'96*, pp.45-46, 1996.

PERFORMANCE COMPARISON BETWEEN OC-FD/DS-CDMA AND MC-CDMA OVER RAYLEIGH FADING CHANNELS

J. Lee, R. Tafazolli, and Barry G. Evans

Mobile Communication Research Group
Centre for Communication Systems Research(CCSR)
University of Surrey, Guildford, Surrey GU2 5XH, England
Tel: +44-1483-300800 ext. 3033 Fax: +44-1483-259504
email: J.Lee@ee.surrey.ac.uk

Abstract: Multiple narrower band carrier systems, i.e. *OC-CDMA*, where overlapping of adjacent carriers are permitted, and multi-carrier CDMA are compared in terms of capacity for asynchronous transmission and reverse link channel. Effect of chip waveform, predetection diversity, and degree of overlapping on capacity is investigated. Our qualitative and quantitative study exhibits that 1) narrower-band carriers system(with more carriers) outperforms wider-band carriers system(with fewer carriers) over frequency nonselective channels, and 2)multicarrier CDMA systems exploiting frequency diversity give better performance than systems with RAKE reception exploiting multipath diversity over highly frequency selective channel.

K. Fazel and G.P. Fettweis (eds.), Multi-Carrier Spread-Spectrum, 235-242.

I. INTRODUCTION

In the past two decades, code division multiple access[1-5] has been regarded as an efficient technique to mitigate jamming, multipath interference, and multiple access interference, and correspondingly to achieve higher capacity. Growing interest is being focused on multicarrier CDMA systems[6-8] since those systems give some advantages over the conventional single carrier CDMA systems.

FD/DS-CDMA system[4,5] with RAKE reception could be a solution in some environments where low complex systems is required, multiple operators share spectrum, or spectrum is not continuous. As the available spectrum is subdivided into more carriers, the capacity gets decreased[4,5]. This loss of capacity introduced by subdividing spectrum is partly compensated by overlapping of adjacent carriers[3]. FD/DS-CDMA with overlapped carriers scheme(OC-CDMA) has been investigated in [1-3]. This paper compares capacity of OC-CDMA with that of the conventional FD/DS-CDMA where overlapping is not permitted.

Kondo in [6] proposed a MC-CDMA model which has strong immunity to narrow-band jamming. Sourour's model in [7] gives higher capacity and hence it is promising for commercial purpose. In this paper, first, we generalise MC-CDMA system based on Sourour's model, and produce a generalised analysis model. This generalised MC-CDMA model eases comparison among various MC-CDMA models. Second, the average bit error rate will be derived. And finally, the capacity for various MC-CDMA and OC-CDMA will be investigated comparatively.

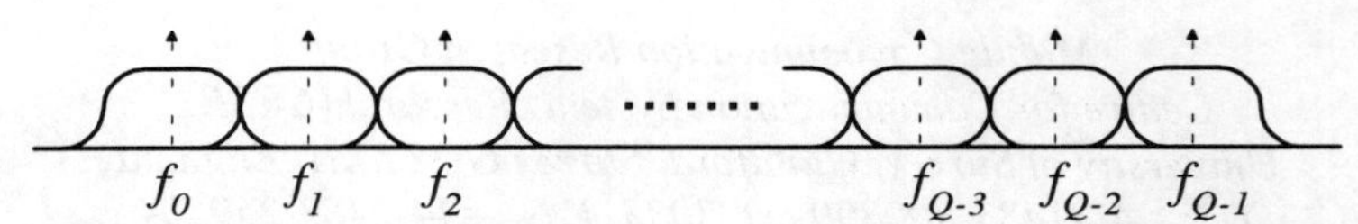

Fig. 1. Power spectral density of the transmitted signal for MC-CDMA.

II. SYSTEM MODELS FOR OC-CDMA AND MULTICARRIER CDMA

A. Generalised MC-CDMA system model

The k-th transmitted signal of a MC-CDMA may be expressed as

$$x^{(k)}(t) = \mathrm{Re}\left\{\sqrt{P^{(k)}}\sum_{m=0}^{M-1}\sum_{s=0}^{S-1} b_{sM+m}^{(k)}(t)\exp\left\{2\pi\left(f_o + f_{sM+m}\right)t + \theta_{sM+m}^{(k)}\right\}\right\} \quad (2\text{-}1)$$

where $P^{(k)}$ denotes a constant value related to the transmission power, and $\theta_q^{(k)}$ (mod 2π) denotes a uniformly distributed random phase of the oscillator. Superscript '$^{(k)}$' denotes the k-th user. Subcarrier centre frequencies f_q's are equally spaced by a certain distance Δ, not necessarily the chip rate in contrast to [7], however any two subcarriers apart by *M-1* subcarriers must satisfy $\left|f_{sM+m} - f_{pM+m}\right| \geq (\Delta f)_c$ where $(\Delta f)_c$ denotes coherence bandwidth of the channel. Fig. 1 illustrates power spectral density of overlapped subcarriers. The baseband signal $b_{sM+m}^{(k)}(t)$, generated by impulse modulator and pulse-shaping filter, may be written in the form

$$b_{sM+m}^{(k)}(t) = \left\{ \sum_{i=-\infty}^{\infty} d_{m,\lfloor i/N \rfloor}^{(k)} a_i^{(k)} w_{sM+m,i} \delta(t - iT_c) \right\} * h(t) . \tag{2-2}$$

The function $\lfloor \cdot \rfloor$ rounds towards minus infinity. $\left(d_{m,i}^{(k)}\right)$ denotes the data stream from the *m*-th branch of serial-to-parallel converter. The signalling interval T is prolonged to M times as long as the original bit duration T_b by serial-to-parallel converter. $\left(d_{m,i}^{(k)}\right)$ and signature sequence $\left(a_i^{(k)}\right)$ generated in every T_c second to combat against multiple access interference are assumed to randomly take values on -1 or +1, and the function $h(t)$ denotes the impulse response of the pulse-shaping filter with its transfer function $H(f)$. Unlikely [7], we assume arbitrary pulse. The second code sequence $\left(w_{q,i}\right)$ for the *q*-th subcarrier is an orthogonal code such as Hadamard code, and removes inter-subchannel interference by establishing orthogonality among subcarriers. Moreover it helps mitigate intermodulation among subcarriers introduced by nonlinear amplifier[10]. The same set of Hadamard code can be used for all users whereas the spreading code $\left(a_i^{(k)}\right)$ must be different for all users. The number of chips per symbol block transmitted simultaneously N is defined by T/T_c and the inverse of chip period $(T_c)^{-1}$ is referred to as the chip rate. In order to exploit frequency diversity instead of multipath diversity, the chipping interval T_c is determined to be longer than maximum rms delay spread T_m, and then only one path per subcarrier is available.

B. OC-CDMA

In OC-CDMA[1-3], the available spectrum is subdivided into multiple subspectra, not necessarily disjoint. Although the carriers are not necessarily with the same bandwidth, we confine our attention to equally wide carriers for simplicity in analysis. Each signal is carried by an element of the carrier set. The k-th transmitted signal in overlapped carrier FD/DS-CDMA(OC-CDMA) system may be expressed as

$$x^{(k)}(t) = \text{Re}\left\{\sqrt{P^{(k)}}\, b^{(k)}(t) \exp\left\{2\pi (f_o + \Delta^{(k)})t + \theta^{(k)}\right\}\right\} \tag{2-3}$$

where $\Delta^{(k)}$ denotes the deviation of carrier centre frequency from the universal centre frequency of available frequency band f_o.

III. CHANNEL MODEL

The channel model in this section is based on the tapped delay line model[9] as follows:

$$c_q^{(k)}(t) = \sum_{l=0}^{L-1} \alpha_{q,l}^{(k)} e^{-j\phi_{q,l}^{(k)}} \delta(t - lT_c) \tag{3-1}$$

where the index q accounts for subcarrier. Constant multipath intensity profile, i.e.

$$E\left\{\left(\alpha_{q,l}\right)^2\right\} = 2\sigma^2 \text{ for any } l \text{ and } q \tag{3-2}$$

is assumed. The parameter L denotes the maximum number of resolvable paths and is given by [9] as

$$L = \left\lfloor \frac{T_m}{T_c} \right\rfloor + 1 \tag{3-3}$$

where the parameter T_m denotes the maximum rms delay spread.

IV. BIT ERROR RATE FOR MC-CDMA AND OC-CDMA

In our receiver model, the output of the matched-filter is sampled in every chip generating period, simply chip period T_c and then summed to generate the decision variable. By some manipulation, the average signal-to-noise ratio of the MRC output has the form

$$\overline{\gamma}_b = S\overline{\gamma}_c \tag{4-1}$$

where

$$\overline{\gamma}_c = \left(\left\{(K-1)[\xi(0) + 2\xi(\Delta)]\right\}\frac{1}{N} + \left(\frac{\overline{E}_b}{SN_o}\right)^{-1}\right)^{-1} \quad \text{for MC-CDMA} \tag{4-2}$$

$$\overline{\gamma}_c = \left(\left\{\frac{L-1}{L} + (K-1)[\xi(\Delta) + 2\xi(\Delta)]\right\}\frac{L}{N} + \left(\frac{\overline{E}_b}{LN_o}\right)^{-1}\right)^{-1} \quad \text{for OC-CDMA} \tag{4-3}$$

with

$$\overline{E}_b = S\sigma^2 PN \int_{-\infty}^{\infty} |H(f)|^2 df, \tag{4-4}$$

and

$$\xi(\Delta) = \frac{1}{T_c} \frac{\sum_{i=1}^{\lceil Q/2 \rceil} \int_{-\infty}^{\infty} |H(f)|^2 |H(f+i\cdot\Delta)|^2 df}{\left(\int_{-\infty}^{\infty} |H(f)|^2 df\right)^2}. \tag{4-5}$$

The function $\lceil \cdot \rceil$ rounds towards plus infinity. The function $\xi(\Delta)$ means the proportionate coefficient of other user interference. The function $\xi(\Delta)$ given by (4-5) is for the midband subcarrier which has maximum average adjacent carrier interference, and hence the average SNR per path given by (4-2) could be regarded as lower bound.

From [9] it is known that for the S-th order of MRC, the BER is related to the average signal-to-noise ratio per path $\bar{\gamma}_c$ as follows:

$$P_e = \left(\frac{1-\mu}{2}\right)^S \sum_{k=0}^{S-1} \binom{S-1+k}{k} \left(\frac{1+\mu}{2}\right)^k \tag{4-6}$$

where $\mu \equiv \sqrt{\frac{\bar{\gamma}_c}{2+\bar{\gamma}_c}}$. Golay(23, 12) code is employed for forward error control, which have error correcting capability t=3. The well known coded BER taking FEC into account is given by [11]

$$P_b \approx \frac{1}{n} \sum_{i=t+1}^{n} i \binom{n}{i} P_e^i (1-P_e)^{n-i} \tag{4-7}$$

where t denotes error correcting capability.

V. Numerical Results

For fair comparison, the average bit energy $\overline{E}_b$ is kept constant at 16dB throughout our numerical results. In Fig. 2, capacity of OC-CDMA for frequency selective and frequency nonselective channels is plotted. In Fig. 2-(a), only one path is available, and the signal suffers from Rayleigh fading. Overlapping increases capacity significantly over frequency nonselective fading channels. Even for smooth rolloff cases, the capacity approaches that of zero rolloff case as spectrum is subdivided further. In contrast, in Fig. 2-(b), the signal is subject to Rayleigh fading as in Fig. 2-(a), but multiple paths are available for diversity. Hence the maximum number of users is large in comparison with that for Fig. 2-(a) since without diversity, the radio system is vulnerable to fading. For smooth rolloff, capacity is slightly increased for small number of wider carriers, but it tends to decrease as spectrum is subdivided further since available paths gets fewer. However, large gain is achievable by overlapping in comparison with nonoverlapping case. For frequency nonselective case, Sourour[7] suggested a solution to increase processing gain: Set S to 1 and increase M towards infinity in order for processing gain to approach that of zero rolloff. And then the capacity for smooth rolloff approaches that of the case with zero rolloff.

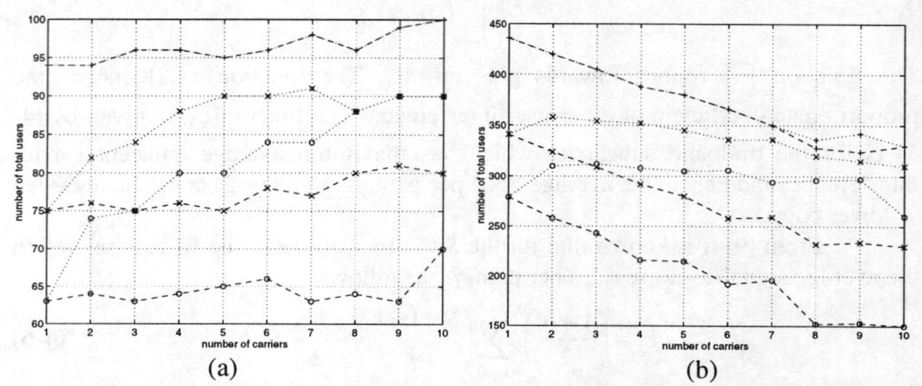

Fig. 2. Maximum number of users in OC-CDMA against the number of carriers for the BER threshold of 10^{-3} with Golay (23,12), available bandwidth 12.5MHz, and uncoded data rate 8kbps(dashed line: no overlapping, dotted line: OC-CDMA, o: β=1, x: β=0.4, +: β=0). (a) Frequency nonselective regardless of carrier bandwidth. (b) Maximum delay spread 2.5μs.

In Table I, capacity is presented for single carrier, and MC-CDMA with and without overlapping. The capacity is determined as the maximum number of users found by increasing chip interval T_c gradually while $T_m \le T_c$ is met. When overlapping between subcarriers is permitted, it is verified that the capacity of smooth rolloff approaches that of zero rolloff.

TABLE I Capacity for single carrier, and MC-CDMA with and without overlapping maximum(M=1, delay spread 8μs, available bandwidth 2.5MHz, (23,12) Golay code, Δ is normalised to the chip rate)

	β=1.0			β=0.4			β=0.0		
	K	Q	Δ	K	Q	Δ	K	Q	Δ
Single carrier	53	1	n/a	65	1	n/a	85	1	n/a
Nonoverlapping	56	20	n/a	67	19	n/a	85	18	n/a
Overlapping	80	18	1.08	82	19	1.05	85	18	1.0

VI. Effect of Pulse Shape on Capacity in Extreme Case

In this section, the effect of pulse shape on capacity in extreme case of $MS \to \infty$ for MC-CDMA is investigated. Raised cosine pulse shaping, main lobe of sinc function(in frequency domain), and rectangular chip pulse(in time domain) are considered for comparison. Fig. 3 plot $\xi(\Delta)$ against distance between carriers in frequency domain. The spacing between adjacent carriers is chosen around the chip rate(1 in Fig. 3). Likewise in [7], increase of processing gain due to overlapping in comparison with single carrier rectangular chip pulse is given by

$$\Delta PG = \frac{MS \cdot 2 / T_c}{(MS-1)\Delta + (1+\beta)/T_c}.$$

As MS increases towards infinity, ΔPG becomes $\Delta PG = 2/(T_c \cdot \Delta)$. On the other hand, as Δ is decreased towards zero, adjacent carrier interference is increased. When K>>1, the relation (4-2) is dominated by the term $[\xi(0) + 2\xi(\Delta)]\frac{1}{N}$. When Δ is less than the chip rate, $\xi(\Delta)$ is increased by the rough relation $\xi(\Delta) \approx 1 - \Delta / T_c$. Because of the figure 2 in the nominator, the increasing rate of processing gain is smaller than the increasing rate of adjacent carrier interference. In contrast, for Δ >1, there is waste of processing gain. Hence Δ =1 is roughly optimal point in terms of capacity. After some manipulation, it follows that $\xi(0) + 2\xi\left(T_c^{-1}\right) = 1$ for all waveforms being considered. In other words, the capacity of MC-CDMA is not dependent on pulse shape for typical waveforms in extreme case.

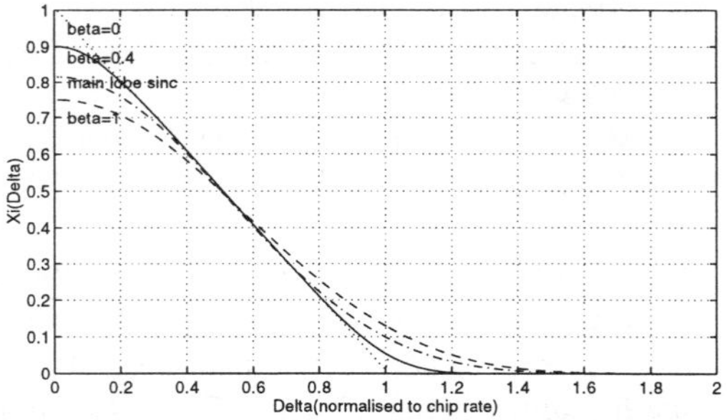

Fig. 3. $\xi(\Delta)$ against Δ(dashed line: β=1, solid line: β=0.4, dotted line: β=0, dash-dotted line: main lobe of sinc function).

VII. CONCLUSIONS

Capacity of MC-CDMA and OC-CDMA has been treated for reverse link and asynchronous transmission. Capacity gain by overlapping of adjacent carriers stems primarily from increased processing gain, and increased order of diversity. Over frequency nonselective channels, OC-CDMA is a better solution than the conventional FD/CDMA without overlapping. In contrast, over frequency selective channels as spectrum is subdivided into more carriers, subdivision of carriers gives lower capacity since diversity order is reduced. For smooth rolloff spectrum, the processing gain approaches that of zero rolloff by MC-CDMA with large number of subcarriers. However, as nonlinearity is taken into account, the performance of multicarrier system would be significantly deteriorated.

For some typical pulse shape such as rectangular chip pulse(in time domain), main lobe of sinc function(in frequency domain), and raised cosine pulse shaping filter, the capacity of MC-CDMA is independent of pulse shape as the number of subcarriers increases towards infinity. When rectangular pulse is chosen, the systems is easily implemented by FFT and IFFT technique. In contrast, single carrier system, the pulse shape affects the performance significantly.

REFERENCES

[1] E. S. Sousa, "The performance of a direct sequence spread spectrum CDMA system utilizing unequal carrier frequencies," *Proc. IEEE Int. Symposium on Spread Spectrum Techniques and Applications, Yokohama, Japan, 1992,* pp. 303-307.

[2] F. Behbahani, and H. Hashemi, "On spectral efficiency of CDMA mobile radio systems," *Proc. 1994 IEEE Int. Conf. on Communications, New Oreleans, USA*, pp. 505-509, 1994.

[3] J. Lee, R. Tafazolli, and B. G. Evans, "Capacity of the overlapped carriers scheme in FD/DS-CDMA," *Proc. IEEE Int. Symposium on Spread Spectrum Techniques and Applications, Mainz, Germany, 1996,* pp. 375-379.

[4] T. Eng, and L. B. Milstein, "Comparison of hybrid FDMA/CDMA systems in frequency selective Rayleigh fading," *IEEE J. Select. Areas Commun.*, vol. 12, no. 5, pp. 938-951, May 1994.

[5] T. Eng, and L. B. Milstein, "Coherent DS-CDMA performance in Nakagami multipath fading," *IEEE Trans. Commun.*, vol. 43, no. 2/3/4, pp. 1134-1143, Feb/Mar/Apr 1995.

[6] S. Kondo, and L. B. Milstein, "Performance of multicarrier DS CDMA systems," *IEEE Trans. Commun.*, vol. 44, no. 2, pp. 238-246, Feb. 1996.

[7] E. A. Sourour, and M. Nakagawa, "Performance of orthogonal multicarrier CDMA in a multipath fading channel," *IEEE Trans. Commun.*, vol. 44, no. 3, pp. 356-367, Mar. 1996.

[8] R. Prasad, and S. Hara, "An overview of multi-carrier CDMA," *Proc. IEEE Int. Symposium on Spread Spectrum Techniques and Applications, Mainz, Germany, 1996,* pp. 107-114.

[9] J. G. Proakis, *Digital Communications, 3rd Ed.* New York, McGraw-Hill, 1995.

[10]E. Sourour, "The effect of cellular portable unit amplifier non-linearity on the performance of the multi-carrier CDMA system," *Proc. IEEE Int. Symposium on Spread Spectrum Techniques and Applications, Mainz, Germany, 1996,* pp. 1351-1355.

[11]J. Wang, and M. Moeneclaey, "Hybrid DS/SFH-SSMA with predetection diversity and coding over indoor radio multipath rician-fading channels," *IEEE Trans. Commun.*, vol. 40, no. 10, pp. 1654-1662, Oct 1992.